T0372002

Practical Spectroscopy in Agriculture and Food Science

Practical Spectroscopy in Agriculture and Food Science

Yuriy I. Posudin

National Agricultural University,
Kiev, Ukraine

Enfield (NH) Jersey Plymouth

CIP data will be provided on request.

SCIENCE PUBLISHERS
An imprint of Edenbridge Ltd., British Isles.
Post Office Box 699
Enfield, New Hampshire 03748
United States of America

Website: *http://www.scipub.net*

sales@scipub.net (marketing department)
editor@scipub.net (editorial department)
info@scipub.net (for all other enquiries)

ISBN 978-1-57808-505-7

Published by Science Publishers, Enfield, NH, USA
An imprint of Edenbridge Ltd.
Printed in India

Preface

The book *Practical Spectroscopy in Agriculture and Food Science* introduces the students and specialists of agricultural and food science to the fundamentals of optical spectroscopy, main principles of modern spectroscopic instrumentation, advantages and practical applications of spectroscopic methods to investigation of agricultural objects such as milk and dairy products, eggs, honey, animal hair, and agronomic plants.

The significant achievements of several years of research activity by myself and my colleagues from The National Agricultural University of Ukraine are reflected in scientific articles which are published in Russian or Ukrainian. The correspondence with many people from various countries who expressed interest in this activity, provided the stimulus to publish a monograph in English, where the experience gained in the field of agricultural spectroscopy can be made available to a wide audience.

The principal sections of the material in this monograph were taught as lecture series by the author who visited the University of Georgia, USA, as a Fulbright scholar in 1996.

The author is grateful to his Ukrainian and Russian colleagues V.I. Kostenko, O.V. Tsyganjuk, P.P. Tsarenko, T.K. Gordienko, S.A. Ivanitskaya, B.M. Gopka, V.F. Lepeshenkov, A.L. Trofimenko, V.P. Polishuk, L.S. Chervinsky, N.I. Yasinetska, T.L. Zharkikh, A.D. Alyokhin, G.I. Gaydidey, M.G. Skrishevskaya, E.M. Galushko, N.G. Golubeva, M.M. Krolivets and V.A. Myasnikov.

Special appreciation is expressed to Professor Stanley J. Kays and Dr. G.G. Dull, University of Georgia, USA, who introduced the author to the fascinating world of non-destructive quality evaluation of agricultural products.

The author is also grateful to Professor Dr. H. Lichtenthaler for being given the opportunity to carry out the study of fluorescence analysis of plants in the Karlsruhe University, Germany.

This book is intended for agricultural and food specialists, teachers, students and postgraduate students.

Yuriy I. Posudin

Contents

CHAPTER

1

Fundamentals of Spectroscopy

1.1 OPTICAL RADIATION

Optical radiation (*light* in wide meaning of the word) includes the electromagnetic waves that have a wavelength from about 300 nm (ultraviolet region), through the visible region (400-700 nm, that can be perceived by human vision), and into the infrared region (from 760 nm to 2,500 nm).

Optical radiation has a *dual nature* because it is found both as waves and particles. According to the classical interpretation, electromagnetic radiation consists of changing electric and magnetic fields that propagate through space forming an electromagnetic wave (Fig. 1.1). This wave is

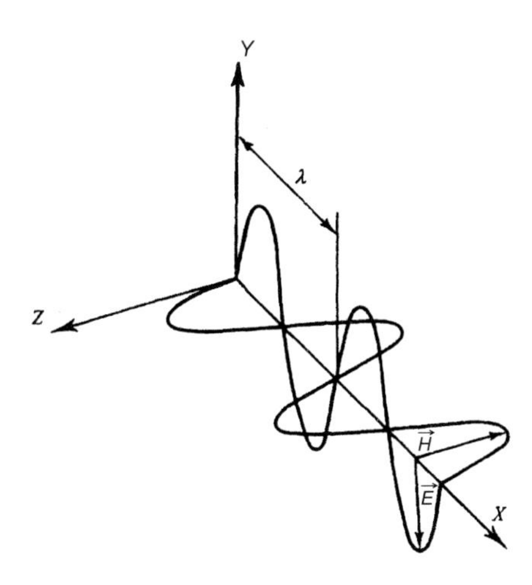

Fig. 1.1 Electromagnetic radiation consists of changing electric and magnetic fields that propagate through space forming an electromagnetic wave

characterized by the amplitude, wavelength and polarization. In terms of the quantum theory, electromagnetic radiation consists of particles called *photons* – discrete packets (*quanta*) of electromagnetic energy that move at the speed of light. Each photon is characterized with its energy $E = h\nu$ and momentum $p = h/\lambda$, where h is Planck's constant, ν is the frequency, and λ is the wavelength of the electromagnetic wave. Dualism of light is presented by the formula $E = h\nu$, where the energy E is related to the particle, and frequency ν is associated with the wave. That is why photon is depicted graphically as a wavy arrow (Fig. 1.2).

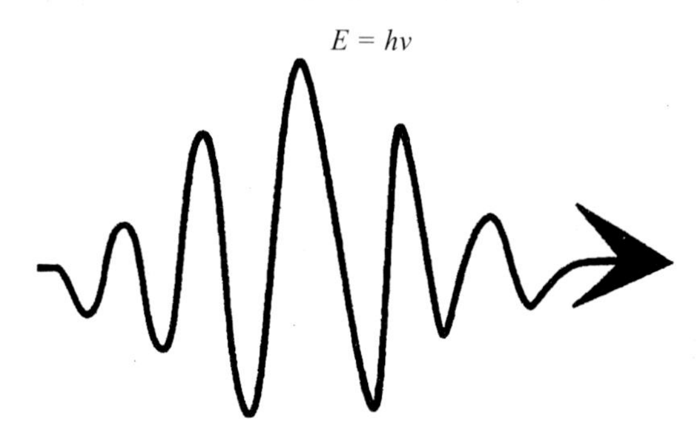

Fig. 1.2 Graphical depiction of photon as a wavy arrow

1.2 INTERACTION OF LIGHT AND AGRICULTURAL OBJECTS

An optical radiation that interacts with agricultural objects takes part in some processes: it can be absorbed by this object and transmitted by it, reflected by its surface, scattered by structural elements, and re-emitted by internal pigments (Fig. 1.3) [Ingle and Crouch, 1988; Analytical Instrumentation Handbook, 1997]. If we neglect the participation of scattering and re-emission, this interaction can be described by the following equation:

$$I_0 = I_a + I_t + I_r \quad \text{(Eqn. 1.1)}$$

where I_0 is the total intensity of radiation incident on the object; I_a, I_t, and I_r describe the intensity of radiation absorbed, transmitted and reflected correspondingly by the object.

Dividing the equation (1.1) on I_0, it is possible to obtain the following expression:

$$1 = \frac{I_a}{I_0} + \frac{I_t}{I_0} + \frac{I_r}{I_0} = \alpha + \tau + \rho \quad \text{(Eqn. 1.2)}$$

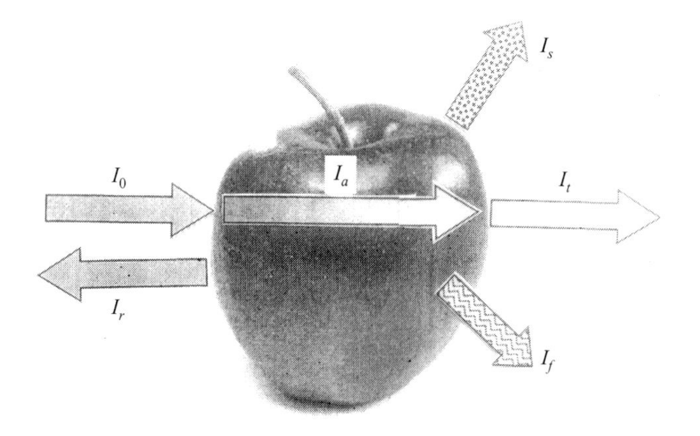

Fig. 1.3 An interaction of optical radiation I_0 with agricultural objects: absorption I_a and transmission I_t of it by this object, reflection I_r by its surface, scattering I_s by structural elements, re-emission I_f by internal pigments

where $\alpha = \frac{I_a}{I_0}$ is the *coefficient of absorption* (or *absorptance*), $\tau = \frac{I_t}{I_0}$ – *coefficient of transmission* (or *transmittance*), and $\rho = \frac{I_r}{I_0}$ – *coefficient of reflection* (or *reflectance*).

1.3 QUANTUM THEORY

The principal ideas of quantum theory were developed over a period of time from 1900 till 1930. According to Planck's hypothesis, absorption and emission of light by atoms and molecules occurs not continuously but by discrete portions – *quanta*, the energy of which is equal:

$$E = h\nu \qquad \text{(Eqn. 1.3)}$$

where $h = 6.63 \cdot 10^{-34}$ J·s – Planck's constant, ν – frequency.

Quantum of electromagnetic radiation (in a strict sense – of light) is called *photon* (from Greek *phos, photos* – light).

In 1913, N. Bohr postulated that atoms and molecules have only certain stationary states which correspond to the discrete number of allowed values of energy $E_i (i = 1, 2, 3...)$. The change of this energy is related to quantum (intermittent) transition from one stationary state to another (Fig. 1.4*a*); it is impossible to occupy the intermediate state between stationary states (Fig. 1.4*b*).

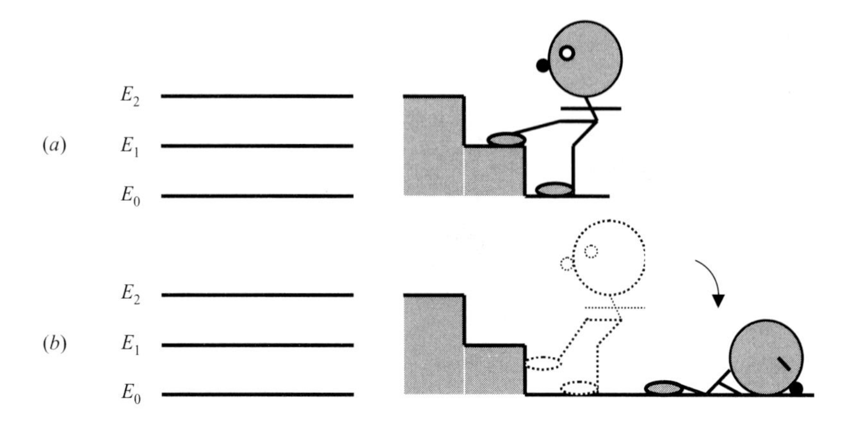

Fig. 1.4 The change of energy is related to quantum (intermittent) transition from one stationary state to another (*a*); it is impossible to occupy the intermediate state between stationary states (*b*)

1.4 ATOMIC ABSORPTION AND EMISSION

Let us discuss the two-level energy structure of an atom. It consists of *ground* level E_0 and *excited* level E_1. Absorption or emission of photon takes place if the energy of photon is equal exactly to the energy difference between state with energy E_i and state with energy E_j:

$$E_i - E_j = h\nu \quad \text{(Eqn. 1.4)}$$

Absorption of photon with energy $h\nu = E_1 - E_0$ by atom promotes it from the ground state to excited state (Fig. 1.5). Atoms are in excited state about 10^{-8} s, and than they relax to their ground state. A downward transition is accompanied with emission of photon of energy $h\nu = E_1 - E_0$ (Fig. 1.6). A plot of the decrease in radiant power during absorption (Fig. 1.7*a*) or emission (Fig. 1.7*b*) as a function of frequency (Fig. 1.7*c*) or wavelength (Fig. 1.7*d*) is called *absorption* (or *emission*) *line.*

1.5 MOLECULAR ABSORPTION AND EMISSION

A molecule is quite a complex system which consists of a number of atoms that participate in three types of motion: *electronic* (movement of electrons around nuclei), *vibrational* (oscillation of nuclei around the equilibrium state), and *rotational* (rotation of the molecule as a whole system in space). That is why the energy E of a molecule is associated with the corresponding types of energy:

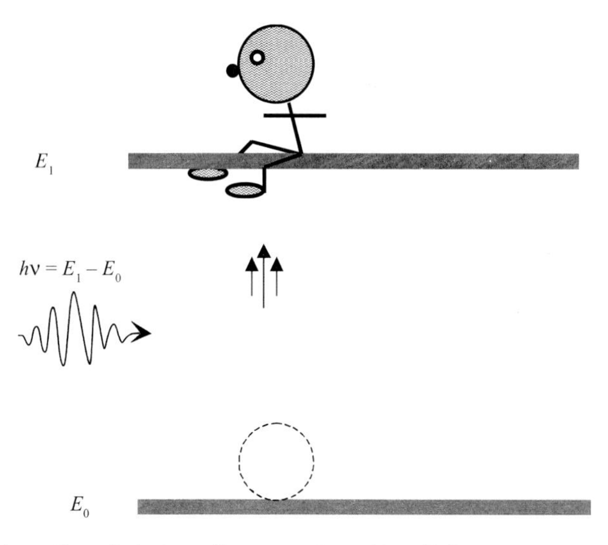

Fig. 1.5 Absorption of photon with energy $h\nu = E_1 - E_0$ by atom promotes it from the ground state to excited state

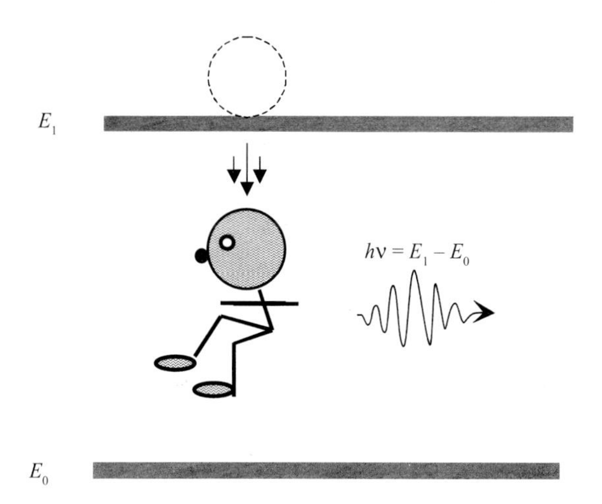

Fig. 1.6 A downward transition is accompanied with emission of photon of energy $h\nu = E_1 - E_0$

$$E = E_{el} + E_{vibr} + E_{rot} \quad \text{(Eqn. 1.5)}$$

where E_{el} is the electronic energy of molecule, E_{vibr} – the energy associated with vibrations of atoms in the molecule, E_{rot} – the energy due to rotational motions of atoms within the molecule.

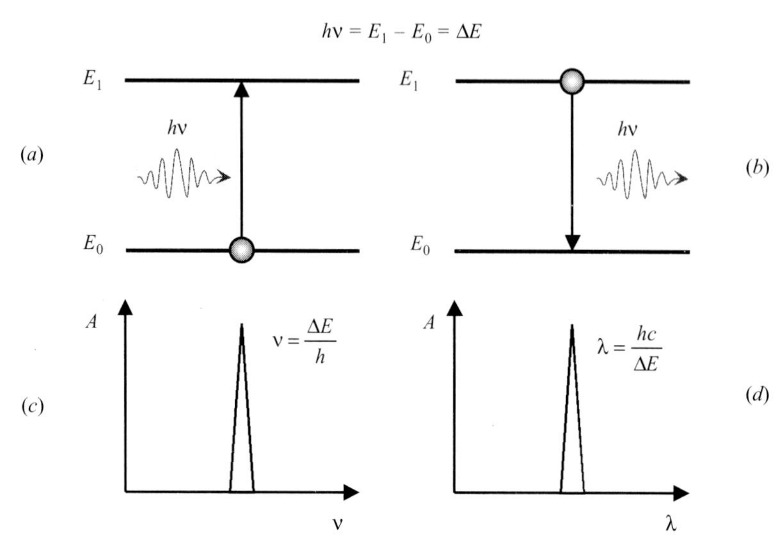

Fig. 1.7 A plot of the decrease in radiant power during absorption (*a*) or emission (*b*) as a function of frequency (*c*) or wavelength (*d*) is called absorption (or emission) line

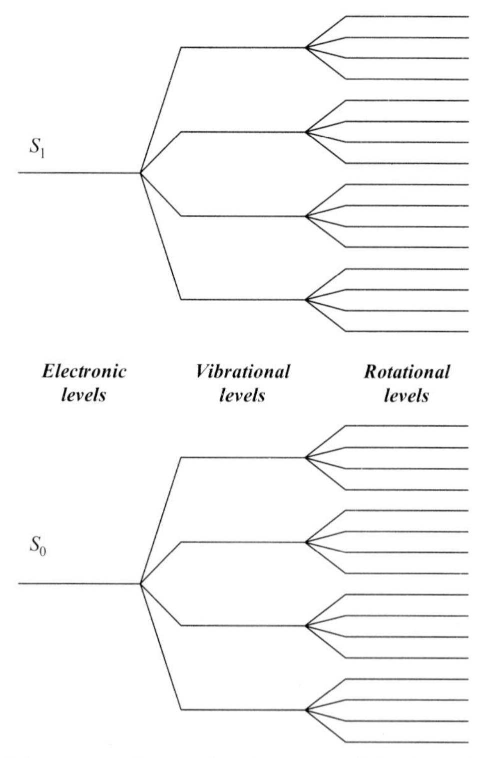

Fig. 1.8 The total energy of a molecule has certain discrete values which are characterized by a set of electronic, vibrational and rotational levels

The total energy E has certain discrete values which are characterized by a set of electronic, vibrational and rotational levels (Fig. 1.8). The transitions between these levels are accompanied by the formation of broad bands of absorption (Fig. 1.9).

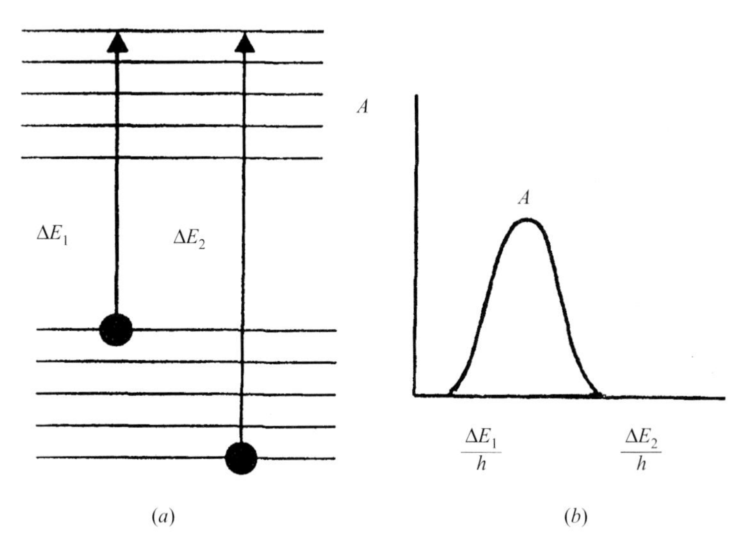

Fig. 1.9 The transitions between electronic, vibrational and rotational levels are accompanied by the formation of broad bands of absorption

The process of molecular emission is characterized by the transition of excited molecules between sublevels; the maximum of emission band is shifted to the long-wave region in comparison with the maximum of absorption band (Fig. 1.10).

1.6 ELECTRONIC ENERGY LEVELS

The fact is that each electron of molecule moves not only in orbital motion, but also rotates around its own axis. The energy of rotation is quantized also. The angular momentum of electron is called *spin*. According to the *Pauli Exclusion Principle* no two electrons in an atom can fit in an orbital; furthermore, the two must have opposed spin states. A molecular electronic state in which all electron spins have opposed electron spins is called a *singlet* state. If the two electrons are parallel, such a molecular state is called a *triplet* state. The average lifetime of an excited singlet state is 10^{-5} - 10^{-8} s, and that of an excited triplet state - from 10^{-4} s to several seconds. The whole set of energy levels of the molecule is called *energy level diagram* (Fig. 1.11).

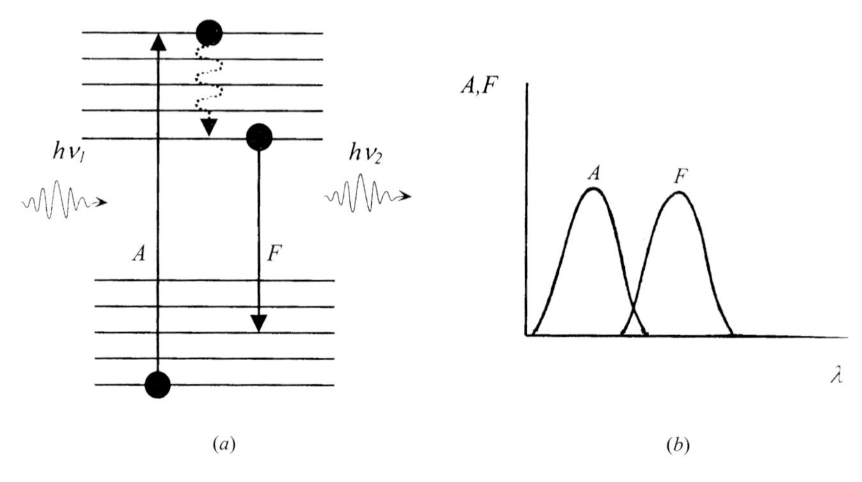

Fig. 1.10 The maximum of emission band is shifted to the long-wave region in comparison with the maximum of absorption band

1.7 VIBRATIONAL ENERGY LEVELS

A molecule is a flexible structure whose atoms participate in vibrational motion. Vibrational energy of such a system is quantized. The allowed energies of vibration are given by:

$$E_{vibr} = (v + ½)h\nu \qquad \text{(Eqn. 1.6)}$$

where v = 0, 1, 2,... – *vibrational quantum number.*

Quantum theory states the following selection rules: the only allowed vibrational transitions are those in which $\Delta v = \pm 1$. The transition from energy state 0 to 1 for any of the vibrational states (v_1, v_2, v_3, ...) is considered a *fundamental*, if $\Delta v = \pm 1$. The transition from ground energy state to a state v_i = 2, 3, ..., is called an *overtone*. Transitions from ground energy state to a state v_i = 1 and v_j = 1 simultaneously are known as *combinations* [Handbook of Near-Infrared Analysis, 1992].

1.8 PROCESSES OF DEACTIVATION

Molecules in excited state can pass to a lower-energy state through *nonradiative* and *radiative* processes [Ingle and Crouch, 1988; Analytical Instrumentation Handbook, 1997].

Nonradiative processes involve:

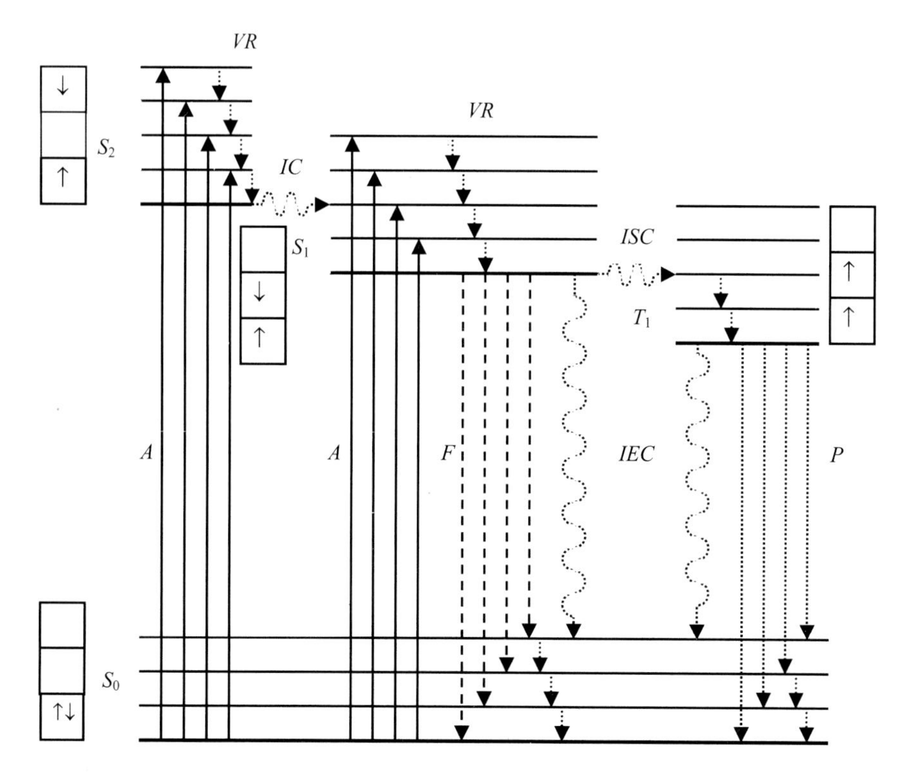

Fig. 1.11 Energy level diagram as the whole set of energy levels of molecule. *A* – absorption, *F* – fluorescence, *P* – phosphorescence, *VR* – vibrational relaxation, *IC* – internal conversion, *ISC* – intersystem crossing, *IEC* – internal and external conversion, S_0 – ground electronic state, S_1 – first singlet electronic state, S_2 – second singlet electronic state, T_1 – first triplet electronic state

Vibrational relaxation VR accompanies the transition of molecule to the ground vibrational state in a given electronic excited state; the energy of the molecule is transferred into heat or vibrational motion. A typical period of vibrational relaxation is 10^{-14} - 10^{-12} s.

Internal conversion IC appears when lower vibrational levels of the upper electronic state overlap upper vibrational levels of lower electronic state. This process lasts for 10^{-12} s.

Intersystem crossing ISC takes place when the vibrational states of singlet and triplet electronic excited states are overlapped.

Radiative processes involve:

Fluorescence F means a radiational transition between ground vibrational state of singlet excited electronic state S_1 and various vibrational states of ground electronic state S_0:

$$S_1 \rightarrow S_0 + h\nu \qquad \text{(Eqn. 1.7)}$$

Fluorescence requires 10^{-10} s to take place.

Phosphorescence P refers to radiational transitions between ground vibrational state of triplet excited electronic state T_1 and various vibrational states of ground electronic state S_0:

$$T_1 \rightarrow S_0 + h\nu \qquad \text{(Eqn. 1.8)}$$

Usually, this process lasts 10^{-10} - 10^4 s.

All these processes of deactivation are depicted in Fig. 1.11 by wave arrows (nonradiative processes) and straight vertical arrows (radiative processes).

1.9 SPECTROSCOPY

Spectroscopy is a scientific study of spectra of electromagnetic radiation for qualitative and quantitative analysis of the structure and properties of matter.

A *spectrum* is a dependence of the intensity of radiation absorbed, transmitted, reflected, scattered or re-emitted by a sample on the wavelength or frequency of electromagnetic radiation.

The following types of spectroscopic methods were used during the investigations of agricultural objects [Posudin, 1996]: *Visible Absorption/Transmission Spectroscopy; Visible Reflectance Spectroscopy; Infrared Spectrophotometry; Near-Infrared Reflectance Spectroscopy; Fluorescence Spectroscopy; Light Scattering Technique.*

The main agricultural objects such as milk, eggs, honey, animal hair and agronomic plants were used for spectroscopic investigation [Posudin, 1988].

1.10 SUMMARY

Investigation of the dependence of the intensity of radiation absorbed, transmitted, reflected, scattered or re-emitted by an agricultural object (milk, eggs, honey, animal hair and agronomic plants) on the wavelength or frequency of optical radiation provides qualitative and quantitative analysis of the structure and properties of these objects.

1.11 REFERENCES

Burns D.A., Ciurczak E.W., eds. 1992. *Handbook of Near-Infrared Analysis.* Marcel Dekker, Inc, New York-Basel-Hong Kong. p 681.

Galen Wood Ewing, ed. 1997. *Analytical Instrumentation Handbook.* Marcel Dekker, Inc., New York-Basel. p 1453.

Ingle, J.D. and Crouch, S.R. 1988. *Spectrochemical Analysis.* Prentice Hall, Englewood Cliffs, New Jersey. p 590.

Posudin, Yu. I. 1988. *Spectroscopy Monitoring of Agrosphere.* Urozhaj, Kyiv. p 128.

Posudin, Yu. I. 1996. *Methods of Optical and Laser Spectroscopy in Agriculture.* Seminar Series of visiting Fulbright Scholar. University of Georgia, U.S.A. January-April 1996.

Posudin, Yu. I. 2005. *Methods of Nondestructive Quality Evalutation of Agricultural and Food Products.* Aristey, Kyiv. p 408.

CHAPTER

2

Laser: Principles and Mechanisms of Action

2.1 PRINCIPLE OF ACTION

Laser is a device which generates coherent electromagnetic waves due to stimulated radiation by an active medium that is placed in an optical resonator. The principle of laser action can be explained by an acronym for **L**ight **A**mplification by **S**timulated **E**mission of **R**adiation (**LASER**).

The main processes of laser action include *pumping, spontaneous emission, stimulated emission* and *absorption* [Svelto, 1976].

2.2 PUMPING

This is a process that provides the excitation of the active particles (atoms, molecules, ions) by means of an external source of energy. These particles are transferred from ground energy level to the higher energy levels during pumping. Some excited states are long-lived; these states are termed as *metastable*.

2.3 SPONTANEOUS EMISSION

This process is realized through the transition of excited particles from the metastable state E_1 to the ground state E_0. Such a radiative process is related to the emission of photon with the energy $h\nu = E_1 - E_0$. It is necessary to mention that all the emitted photons have different phases and directions of propagation (Fig. 2.1). In such a way, spontaneous radiation is *incoherent* monochromatic radiation.

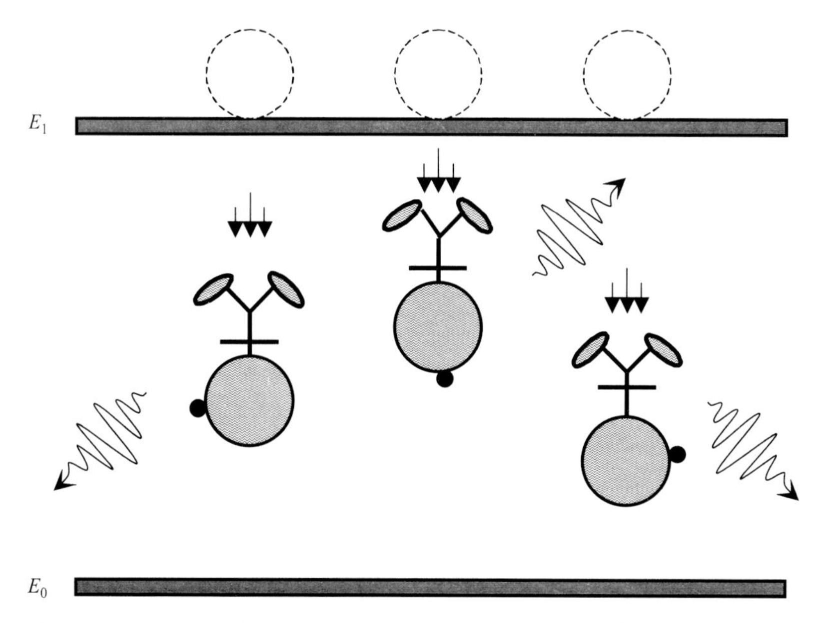

Fig. 2.1 Spontaneous radiation as incoherent monochromatic radiation which consists of the emitted photons that have different phases and directions of propagation

2.4 STIMULATED EMISSION

This kind of radiation takes place when an external photon with energy $h\nu = E_1 - E_0$ strikes the excited particle and provokes the transition of these particles from the excited state to the ground state. The most important feature of this process is that emitted photons travel in exactly the same direction and are precisely in phase with the external photon which caused stimulated emission. That is why stimulated emission is *coherent* with the incoming radiation (Fig. 2.2). As soon as both the incoming and emitted photons travel in phase, there is an *amplification* of the light - a single photon stimulates the appearance of several photons (Fig. 2.3).

2.5 ABSORPTION

There is another process which competes with light amplification; this process is called *absorption*. If the external photon interacts with the laser medium, a certain number of particles are transferred from E_0 to E_1; this process is characterized by decreasing of the number of photons (Fig. 2.4).

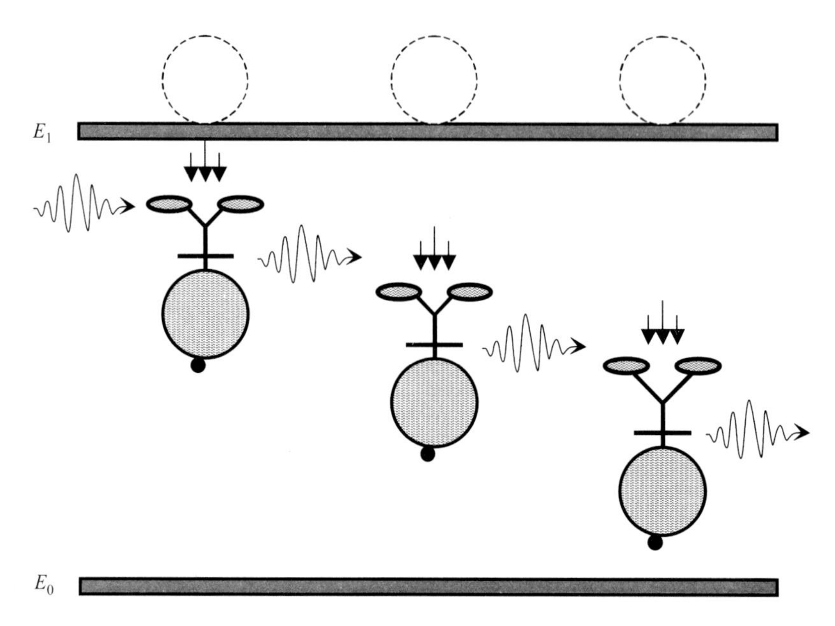

Fig. 2.2 Stimulated emission as coherent monochromatic radiation which consists of the emitted photons that travel in phase

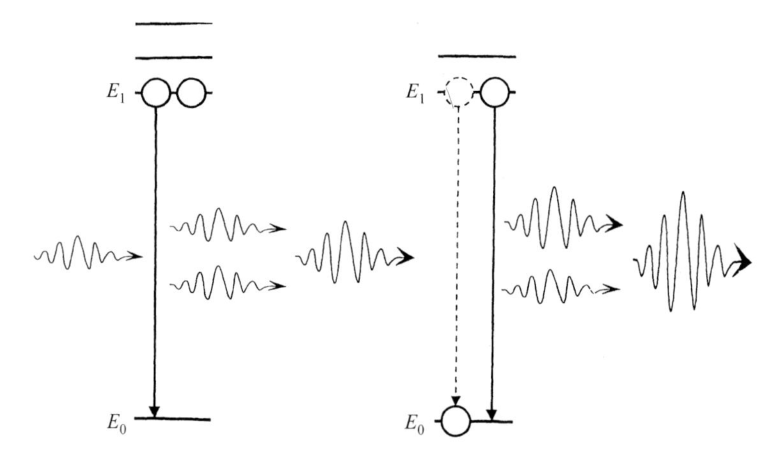

Fig. 2.3 Amplification of the light as a result of travelling of both the incoming and emitted photons in phase, and the appearance of several photons stimulated by a single photon

2.6 INVERSION OF POPULATION

The number of photons produced by stimulated emission must exceed the number of photons lost by absorption in order to provide amplification of light. This condition means that the number of excited

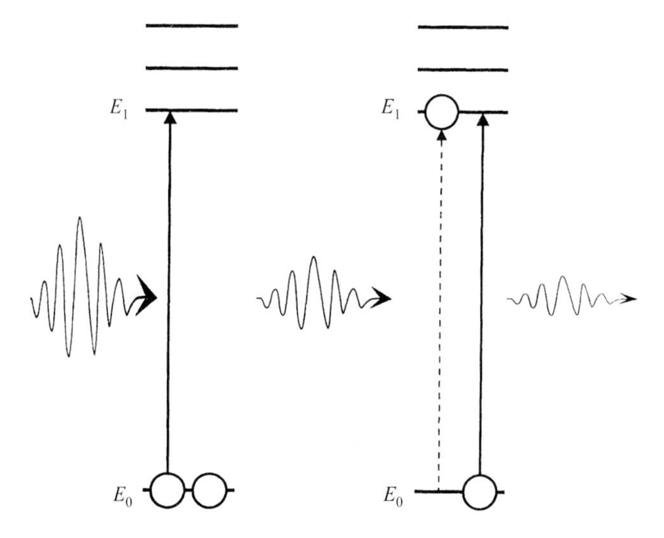

Fig. 2.4 Absorption as the interaction of the external photon with the laser medium; this process induces a transfer of a certain number of particles from E_0 to E_1 and is characterized by decreasing of the number of photons

particles must prevail the number of particles in the ground state. Such a situation can be realized due to pumping which causes *inversion of population* – the redistribution of the particles population between the energy levels (Fig. 2.5).

2.7 COMPONENTS OF LASER

The main components of laser are *active medium* (gas or mixture of gases, solution of organic dye, solid crystal, or semiconductor), *optical resonator* (two flat or concave mirrors, one of which is semitransparent), and *source of pumping* (electrical discharge, electron beam, light flash, gas dynamic system, or chemical reactor). Typical design of laser is given in Fig. 2.6.

2.8 FORMATION OF LASER RADIATION

Let us consider the behaviour of spontaneously emitted photons in the active medium of laser (Fig. 2.7) [Posudin, 1985; 1988; 1989]. Photon *1* propagates at some angle to the longitudinal axis of the resonator which is formed with two mirrors are numbered M_1 and M_2. This photon leaves the resonator and does not take part in laser generation; what happens to it is not dealt with here as it is of no consequence in this context. A similar behaviour is demonstrated by photon *2* after reflection from the

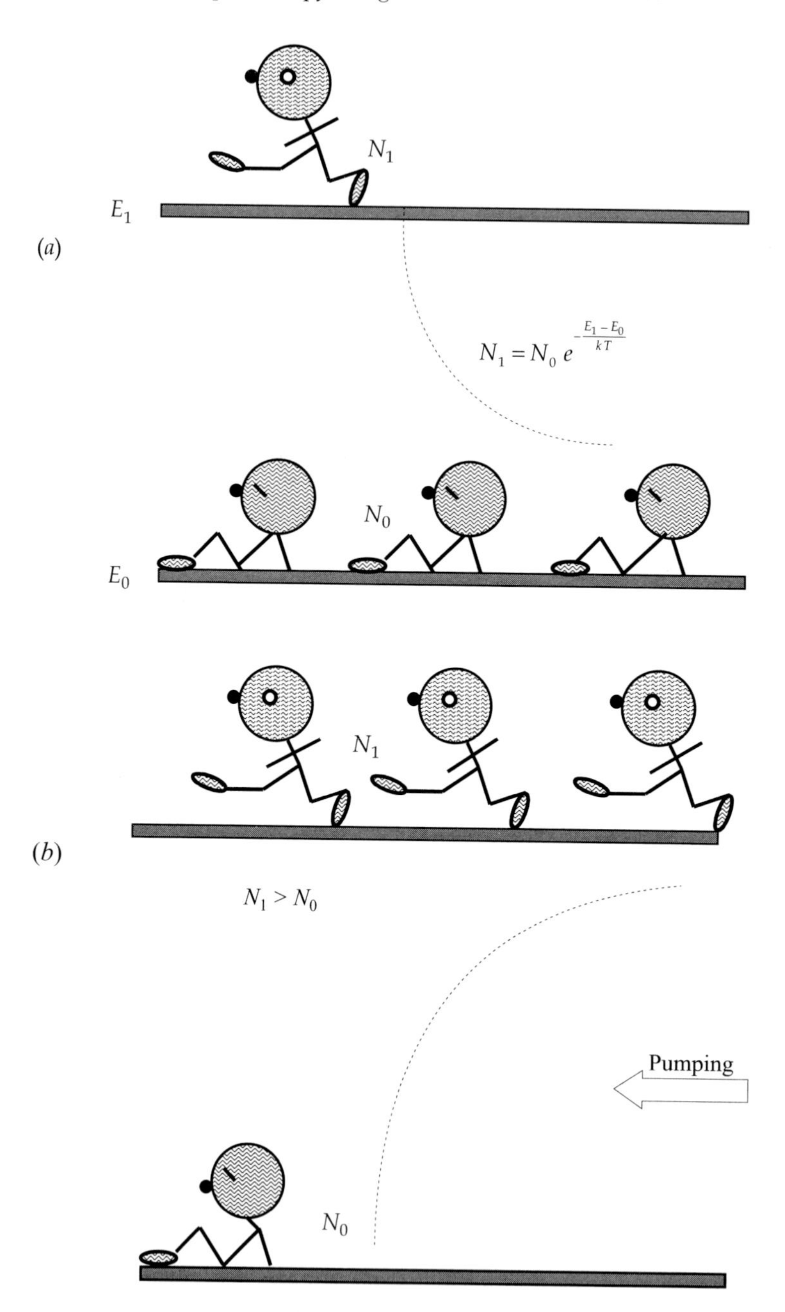

Fig. 2.5 Inversion of population as the redistribution of the particles population between the energy levels (*a*); this condition means that the number of excited particles, N_1 must prevail the number of particles in the ground state, N_0 under the effect of pumping (*b*)

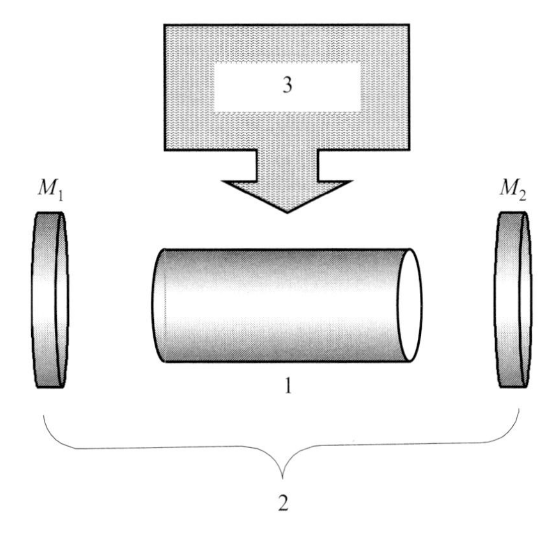

Fig. 2.6 The main components of laser: 1 – active medium (gas or mixture of gases, solution of organic dye, solid crystal, or semiconductor), 2 – optical resonator (two flat or concave mirrors, one of which is semitransparent), and 3 – source of pumping (electrical discharge, electron beam, light flash, gas dynamic system, or chemical reactor)

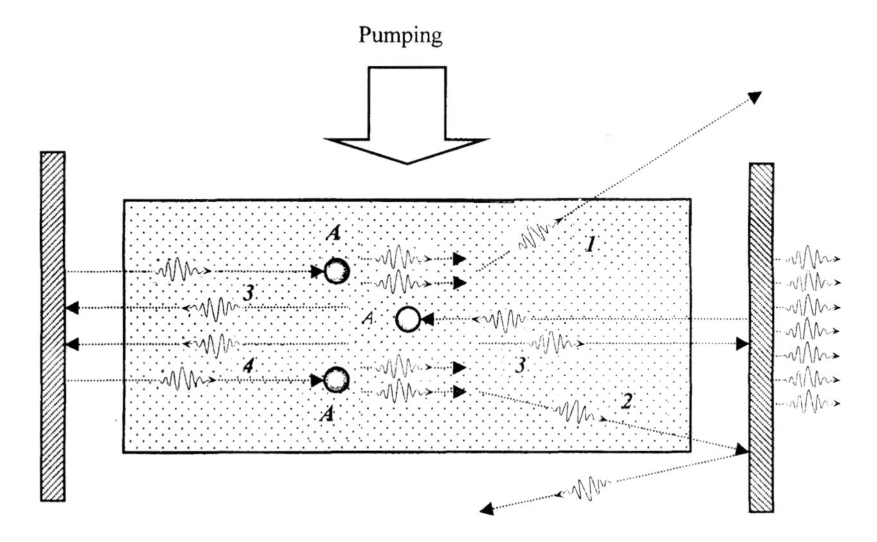

Fig. 2.7 Formation of laser radiation (explanation in text)

surface of the mirror. Photon *3* propagates along the longitudinal axis of the resonator; it reflects from the surface of the mirror, reverses and interacts with the excited particle *A* (atom or molecule) of the active medium. This particle is transferred from the excited level to the ground level and emits a new photon 4 as a result of stimulated emission. Both photons 3 and 4 travel in the same direction. They propagate in parallels to the longitudinal axis of the resonator, reflect from mirrors, and the process of stimulated emission is repeated. One of the mirrors has 100% reflection, and the other is partially transparent. The output laser beam is formed with only those photons whose direction of propagation is in parallels to the longitudinal axis. That is why a laser beam is characterized by a small divergence.

2.9 PROPERTIES OF LASER RADIATION

The main properties of laser radiation are monochromaticity, coherence, directionality, and brightness. Some types of lasers have the possibility to change the frequency (or wavelength) of radiation. The other laser devices produce ultrashort pulses of radiation.

Monochromaticity means a narrow spectral interval of radiation, which can be characterized approximately by one frequency (or wavelength) only.

Coherence is the co-ordinated proceeding of several wave processes in space and time. *Temporal coherence* can be distinguished by the following factor: if for a given point in space there is a constant phase difference between the amplitude of the wave at two successive instances in time; and *spatial coherence* – which is characterized by a constant time-independent phase difference for amplitude at two different points.

Directionality is determined by the angular distribution of the laser beam. This property is the direct result of the propagation of two light waves in the optical cavity; only waves, which propagate normally to the resonator mirrors, participate in the formation of a laser beam. The angular distribution, or *divergence* of gas lasers is about 10′; of solid-state lasers 10′–40′; of semiconductor lasers about 30°.

Brightness of the source of electromagnetic waves is defined as the power of radiation which is emitted from the unit of the source surface through the unit solid angle. As soon as the laser beam has a small divergence, it has a brightness which is several orders higher than the traditional sources of light.

2.10 SUMMARY

Unique properties of laser radiation such as monochromaticity, coherence, directionality, and brightness provide the performance of new nondestructive, fast and precise methods for control of agricultural objects.

The progress in the successful realization of laser methods in agriculture is obvious and many convincing examples of this progress are increasing day to day. The sphere of laser technology in agriculture is open to new investigations and practical applications.

2.11 REFERENCES

Posudin, Yu.I. 1985. *Laser Microfluorometry of Biological Objects*. Vyssha Shkola, Kiev. p 108.

Posudin, Yu.I. 1988. *Lasers in Agriculture*. Science Publishers, Inc., Enfield, NH, USA. p 188.

Posudin, Yu.I. 1989. *Laser Photobiology*. Vyssha Shkola, Kiev. p 246.

Svelto, O. 1976. *Principles of Lasers*. Plenum Press, New York, London. p 373.

CHAPTER

3

Spectroscopic Methods

3.1 ABSORPTION/TRANSMISSION SPECTROSCOPY

All agricultural objects are built out of smaller particles called molecules and atoms. Each atom in a molecule takes part in vibrational movements with a certain vibrational frequency. If a light wave of a certain frequency interacts with an object that has atoms of the same vibrational frequency, then those atoms will absorb the light energy and transfer their vibrational energy into thermal energy through the interaction with neighbouring atoms. The light wave usually consists of many frequencies; the object absorbs certain frequencies selectively. Different types of atoms and molecules have a different vibrational frequency; that is why they selectively absorb different frequencies of light. For example, chlorophyll absorption of blue and red light means the reflection of green light: hence the green colour of all the leaves can be distinguished.

Let us discuss the main terminology which is related to the passing of light through the object (Fig. 3.1).

Absorption is the process in which incident radiated energy is retained without reflection or transmission on passing through a medium.

Absorbance is the ability of a solution or a layer of a substance to absorb radiation that is expressed mathematically as the negative common logarithm of the transmittance of the substance or solution:

$$A = -\lg\left(\frac{I_t}{I_0}\right) = -\lg(1 - \alpha) = -\lg\tau = \alpha Cl \qquad \text{(Eqn. 3.1)}$$

where $\tau = \frac{I_t}{I_0}$ is transmittance, C = concentration, l = thickness of the sample (the path length).

Absorbance is also called *optical density D*, but absorbance is related with absorption inside the substance only, while optical density is determined either by absorption, or by scattering.

Absorptivity (*extinction coefficient*) is used if concentration C is expressed in g/L:

$$a = \frac{A}{Cl} \tag{Eqn. 3.2}$$

The unit of absorptivity is $L \cdot g^{-1} \cdot cm^{-1}$.

Molar absorptivity (*molar extinction coefficient*) is used if concentration is expressed in units of mol/L:

$$\varepsilon = \frac{A}{Cl} \tag{Eqn. 3.3}$$

The unit of molar absorption coefficient is the $L \cdot mol^{-1} \cdot cm^{-1}$.

It is possible to obtain from the equations (2.1) and (2.3) the important relationship that is known as *Beer-Lambert-Bouger Law*:

$$I_t / I_t = e^{-\varepsilon Cl} \tag{Eqn. 3.4}$$

Absorption cross section is used sometimes for the description of the absorption properties of a substance:

$$\sigma = A/(0.434nl) \tag{Eqn. 3.5}$$

where n is the concentration of the substance in atoms or molecules in cm^3.

Fig. 3.1 Interaction of light with the objects is characterized with a rich terminology

Absorption cross section is related to molar absorptivity:

$$\sigma = 3.8 \cdot 10^{-21} \varepsilon \qquad \text{(Eqn. 3.6)}$$

Coefficient of absorption α, absorptivity a, molar absorptivity ε, and cross section σ are related to the absorbance A:

$$A = 0.434\alpha l = 0.434\sigma n l = aCl = \varepsilon Cl \qquad \text{(Eqn. 3.7)}$$

3.2 LIGHT REFLECTION

Reflection is the phenomenon of a propagating light wave being thrown back from a surface. Reflection may be specular or diffuse according to the nature of the interface.

Specular reflection is the reflection of light from a surface where at the point of reflection an incident beam is reflected at (and only at) an angle equal to the angle of incidence (both taken with respect to the perpendicular at that point) (Fig. 3.2*a*).

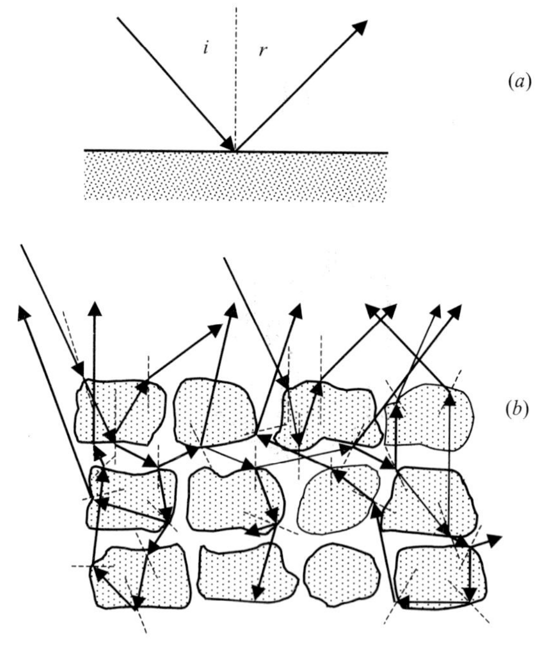

Fig. 3.2 Types of reflection: specular reflection (*a*) provides little or no information about internal properties of an agricultural object in comparison with diffuse reflection (*b*) that is received due to numerous processes of random reflection, refraction, and scattering at interfaces of internal elements inside the sample

Diffuse (or *body*) *reflection* is the reflection of light from an uneven or granular surface such that an incident ray is seemingly reflected at a number of angles; whereas specular reflection provides little or no information about internal properties of an agricultural object in comparison with diffuse reflection that is received due to numerous processes of random reflection, refraction, and scattering at interfaces of internal elements inside the sample (Fig. 3.2*b*).

The following empirical expression can be used for diffuse reflectance R [McClure, 1994]:

$$D = \ln\frac{1}{R} = \ln\frac{I_0}{I_r} \qquad \text{(Eqn. 3.8)}$$

The agricultural objects which absorb weakly optical radiation and contain scattering particles are characterized by the *effect of intensification* of the absorbing bonds: this intensification arises from the radiation traversing an optical path which may be as much as eighty times the sample thickness [Handbook of Near-Infrared Analysis, 1992].

The diffuse reflectance can be estimated in practice through the following equation that demonstrates a nearly linear relationship with concentration [Norris, 1982]:

$$\lg\frac{R'}{R} = \lg\frac{1}{R} + \lg R' \sim aC/s \qquad \text{(Eqn. 3.9)}$$

where R' and R = reflectance of the standard (barium sulphate or magnesium oxide) and of the sample correspondingly, a = constant, C = concentration, s = scattering constant.

For monochromatic radiation equation (3.9) can be rewritten as:

$$C = k + (s/a)\,\lg\left(\frac{1}{R}\right) \qquad \text{(Eqn. 3.10)}$$

where k is absorption constant.

The *Near-Infrared (NIR)* region of the electromagnetic spectrum occupies the wavelength range from 760 nm to 2,500 nm (or from 13,000–4,000 cm^{-1}). All the absorption bands in the NIR region are induced by many vibrational transitions, particularly higher harmonics of fundamental transitions (overtones) and combinations of these harmonics with fundamental transitions. The shape of the NIR spectrum for a given molecule is sharp, unique and characteristic, and can be used for identification and quantitative analysis of the sample. Typical NIR spectra of milk, cream and sour cream are presented in Figs. 3.3–3.5.

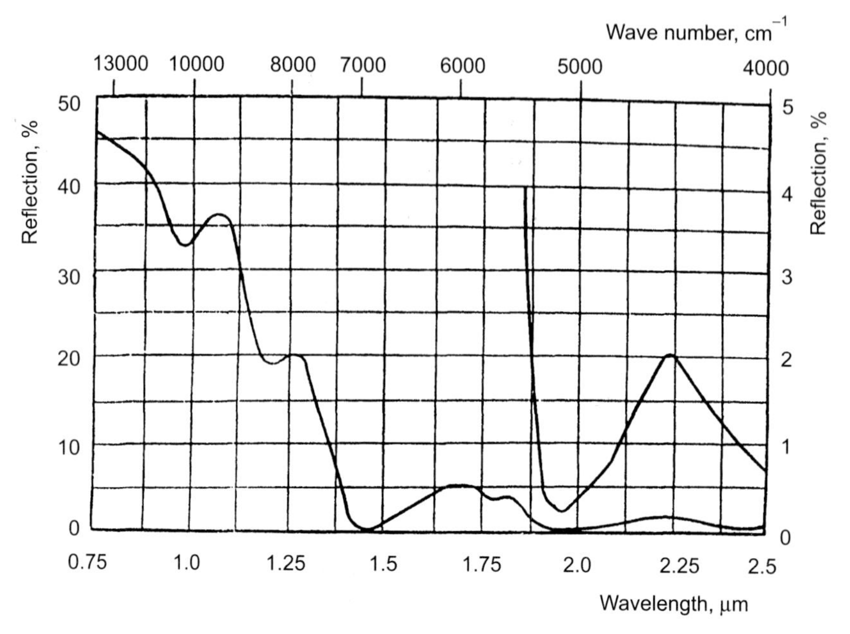

Fig. 3.3 Near-infrared spectra of milk

The presence of characteristic groups such as –CH, –OH, –NH causes the appearance of overtones: thus, fundamental vibrations of CH-group occupy the wavelength range 3,000–3,600 nm; first overtone – 1,600-1,800 nm; second overtone – 1,100–1,200 nm. The region of vibrations of NH-group consists of 2,900 nm (fundamental vibration), 1,500 nm (first overtone), 1,000 nm (second overtone), and 2,200 nm (combination of overtones). The spectral region of OH-group is presented by 2,800 nm (fundamental vibration), 1,400 nm (first overtone), 1,000 nm (second overtone) [Fuhrman, 1987].

In addition, there is participation of water in formation of a NIR spectrum; spectral properties of water are caused by a number of overtones at 760, 970, 1,180, and 12,450 nm, and combination of overtones at 1,940 nm.

3.3 LUMINESCENCE

Luminescence occurs when the energy source transfers an electron of an atom from a lower (ground) energy state into an excited higher energy state, then the electron releases the energy in the form of light when it returns back to a lower energy state. This term refers to emission from

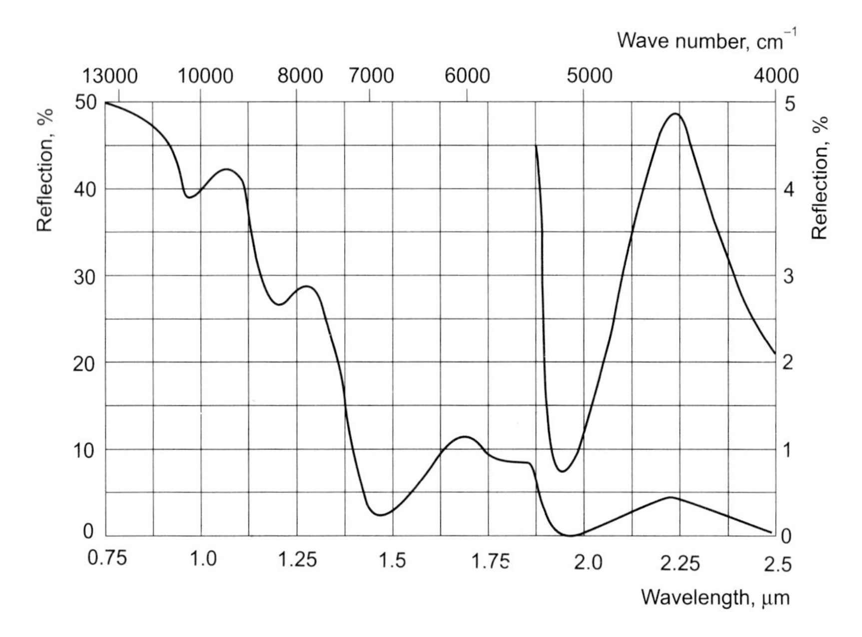

Fig. 3.4 Near-infrared spectra of cream

cool bodies or to emission from hot bodies that is not due to thermal excitation. If the emission of light from excited states is produced by light absorption, such a process is called *photoluminescence*. There are two particular types of photoluminescence - short-lived (10^{-10}–10^{-6} s) *fluorescence* and long-lived (10^{-6}–10^{4} s) *phosphorescence.*

3.4 SCATTERING

Light scattering is the change of some characteristics of the light flow (space distribution of light intensity, frequency spectrum, polarization) during its interaction with the matter. Experiments have indicated that the *elastic scattering* is characterized by the same frequency as that of scattered radiation. This type of light scattering can be classified as *Rayleigh scattering, Debye scattering,* and *Mie scattering*. A concrete type of scattering is determined by the refractive indices of the particle-scatterers and surrounding medium, and by the dimensions of these particles relative to the wavelength of the incident radiation (Table 3.1).

Each type of light scattering is also characterized by a certain angular distribution of scattered radiation intensity: little particles demonstrate the symmetric distribution of scattered light, while large particles

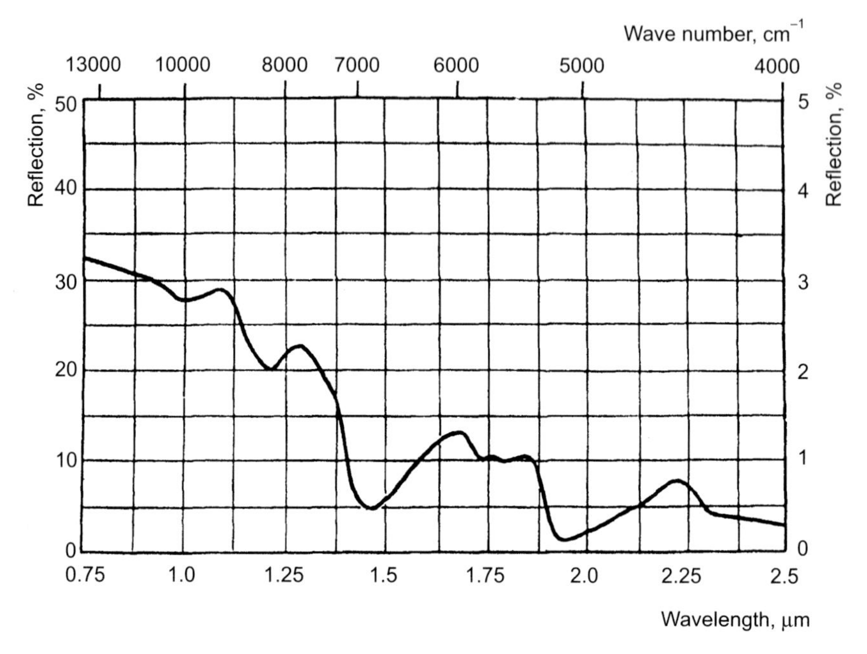

Fig. 3.5 Near-infrared spectra of sour cream

Table 3.1 Types of light scattering [Ingle and Crouch, 1988]

Types of Scattering	*Refractive Index Requirement*	*Size Requirement*
Rayleigh	$\lvert(n-1)\rvert << 1$	$d < 0.05\lambda$
Debye	$\lvert(n-1)\rvert \approx 0.1$	$0.05\lambda < d < \lambda$
Mie	$(n-1) >> 0$	$d > \lambda$

produce the presence of a number of maxima and minima due to interference between the rays reflected from different points of the particle (Fig. 3.6). Large-particle scattering is used in *turbidimetry* and *nephelometry* – methods for determining the amount of cloudiness, or turbidity, in a solution based on measurement of the effect of this turbidity on the transmission and scattering light.

3.5 SUMMARY

Spectroscopic methods are based on the interaction of optical radiation with matter. Such types of interaction can be distinguished as *absorption*

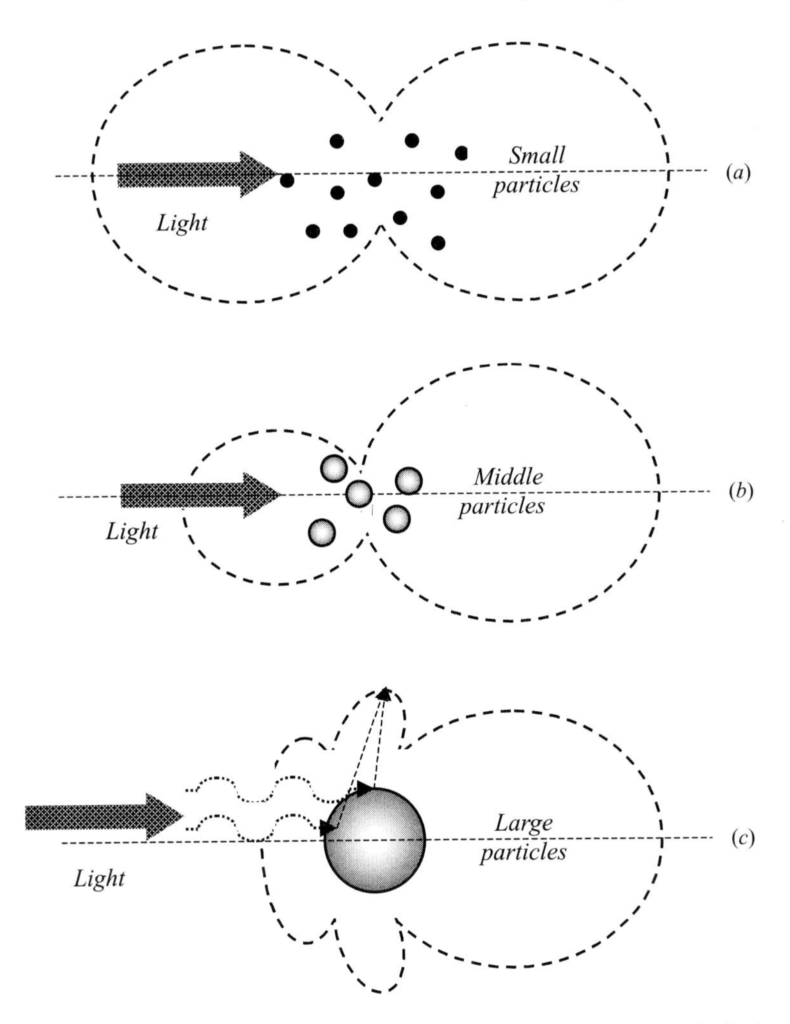

Fig. 3.6 Each type of light scattering is characterized by a certain angular distribution of scattered radiation intensity: *a* – little particles demonstrate the symmetric distribution (Rayleigh scattering); *b* – middle particles produce asymmetric distribution (Debye scattering); *c* – large particles produce the presence of a number of maxima and minima due to interference between the rays reflected from different points of the particle (Mie scattering)

– the process in which incident radiated energy is retained without reflection or transmission on passing through a medium; *reflection* – the phenomenon of a propagating light wave being thrown back from a surface; *luminescence* which occurs when the energy source transfers an electron of an atom from a lower (ground) energy state into an excited higher energy state, then the electron releases the energy in the form of light when it returns back to a lower energy state; *light scattering* – the change of some characteristics of the light flow (space distribution of light intensity, frequency spectrum, polarization) during its interaction with matter. All spectroscopic methods offer non-destructive, rapid and precise evaluation of agricultural products.

3.6 REFERENCES

Burns, D.A. and Ciurczak, E.W., eds. 1992. *Handbook of Near-Infrared Analysis.* Marcel Dekker, Inc., New York-Basel-Hong Kong. p 681.

Fuhrman, I. 1987. Zur anwendung der nah-infrarotreflexions-spektroscopie in der milchwirtschaft. *Lebensmit teilindustrie.*4:171–174.

Galen Wood Ewing, ed. 1997. *Analytical Instrumentation Handbook.* Marcel Dekker, Inc., New York-Basel. p 1453.

Ingle, J.D. and Crouch, S.R. 1988. *Spectrochemical Analysis.* Prentice Hall, Englewood Cliffs, New Jersey. p 590.

McClure, W.F. 1994. Near-Infrared Spectroscopy. In: *Spectroscopic Techniques for Food Analysis,* R.H. Wilson, ed. VCH Publishers, Inc., New York. pp. 13–57.

Norris, K.H. 1982. Multivariate Analysis of New Materials. In: *Chemistry and World Food Supplies: The New Frontiers.* L.W. Shemit, ed. CHEMRAWN II. Pergamon Press, Oxford, England. p. 527.

Posudin, Yu.I. 1996. *Methods of Optical and Laser Spectroscopy in Agriculture.* Seminar Series of Fulbright Scholar. University of Georgia, Athens, U.S.A.

CHAPTER

4

Spectroscopic Analysis of Milk and Dairy Products

4.1 COMPOSITION OF MILK

Milk is a complex mixture of fat, protein, lactose, vitamins, salts and other compounds which can exist in colloidal dispersion or in an aqueous solution of milk components. Approximate composition of milk of European breeds includes 87.00% water, 3.80% non-fat solids, 3.40% protein, 4.50% lactose and 1.30% solids.

Fat of milk is a mixture of glycerides and fatty acids. Fatty acids include polynonsaturated linoleic (two double bonds), linolenic (three double bonds) and arachidonic (four double bonds) acids. Fat globules in milk range from 1 to 10 mm with a density of $(1.5–3.0) \cdot 10^9$ ml^{-1} .

Proteins consist of about 20 amino acids – the most important being tryptophan, tyrosine and phenylalanine. Proteins in milk are presented by casein micelles (0.1–0.2 μm) and serum protein particles (0.01–0.02 μm).

Carbohydrates in milk are present in the form of lactose (disaccharide), glucose and galactose (monosaccharides).

Milk contains *vitamins* such as *A* (retinol), *D* (ergocalcipherol D_2 and cholecalcipherol D_3), *K*, B_1 (thiomin), B_2 (riboflavin), niacin (nicotinic acid), pantothenic acid, B_6, biotin, folic acid, B_{12} (cobalamin), and *C* (ascorbic acid).

Total solids content is the mass remaining after completion of the specified drying procedure and expressed as a percentage of mass.

The *total non-fat solids* means the content of total solids minus the content of fat.

4.2 INFRARED SPECTROPHOTOMETRY OF MILK AND DAIRY PRODUCTS

Spectrophotometry is a study of absorption (transmission) properties of a sample as a function of wavelength.

Infrared absorption of milk is related to the presence of certain structural groups in it [Goulden, 1964; Biggs and Sjaunja, 1987]. Each fat molecule consists of three carbonyl groups (C=O) of triglyceride which are responsible for absorption at 5.73 μm (1,745 cm^{-1}); in addition, carbon–hydrogen groups (CH_2) of triglyceride take part in absorption at 3.4–3.5 μm (2,941–2,857 cm^{-1}). Each molecule of protein consists of amide units which are linked by peptide bonds; the absorption of the amide II band at 6.46 μm (1,548 cm^{-1}) can be used for determination of the total protein content. Each molecule of lactose consists of hydroxyl groups (OH) which absorb at 9.6 μm (1,042 cm^{-1}). Besides, a molecule of water absorbs at 4.3 μm (2,326 cm^{-1}).

4.3 INSTRUMENTATION

4.3.1 Optical Density of Milk

The measurement of milk particles concentration demands the elimination of light scattering – the condition of unitary scattering of optical radiation by milk particles can be performed if:

(1) the value of optical density is equal $D = \lg I/I_0 << 1$;

(2) there is a linear dependence of optical density on the concentration of milk particles $D = f(C)$, if $D << 1$.

The dependence of I/I_0 and $\lg I/I_0$ on the level of dilution of milk sample was measured with the photocalorimeter KFK-2MP in spectral region 315–980 nm. The results are shown in Figs. 4.1*a* and *b*. It is clear that the dependence $D = f(C)$ demonstrates a linear character for diluted samples of milk.

The relation of optical density of milk and wavelength of optical radiation is presented in Fig. 4.2. The spectral curve has a maximum near 590–600 nm which is probably the result of competition of two processes – scattering by milk particles and natural absorption of milk.

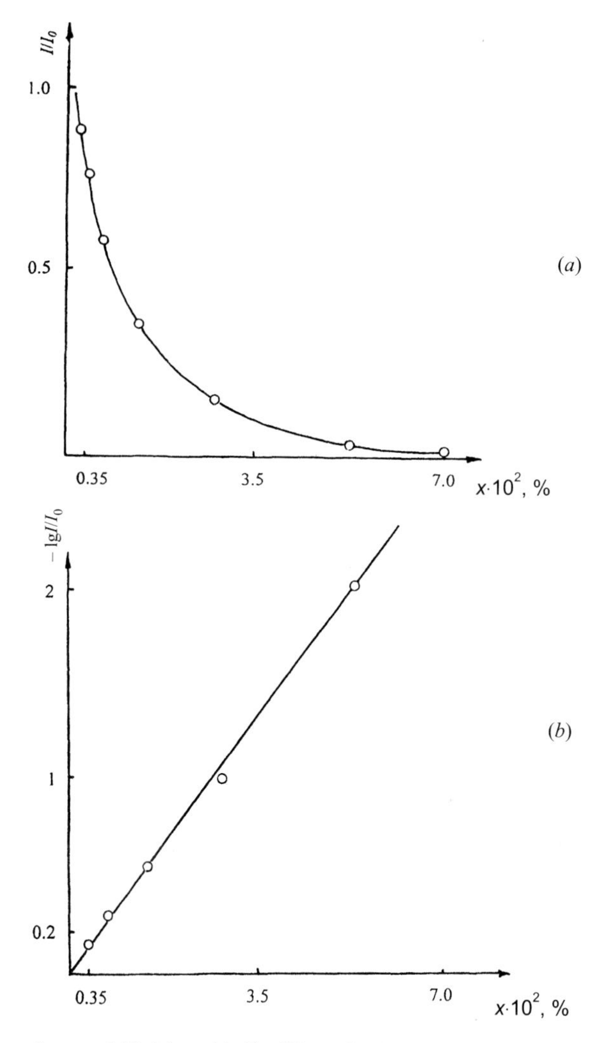

Fig. 4.1 Dependence of I/I_0 (*a*) and $\lg I/I_0$ (*b*) on the level of dilution of milk sample

The main conclusion of this investigation is as follows: a milk sample should be diluted adequately to obtain the linear dependence of optical density on the concentration of milk sample which is to be evaluated. This is a serious limitation to the method of absorption/transmission of a spectroscopy in the visible part of the spectrum.

4.3.2 Infrared Spectrophotometry of Milk and Dairy Products

The main objective of this investigation is to study the characteristic infrared spectra of milk, boiled butter, casein, and lactose. An infrared

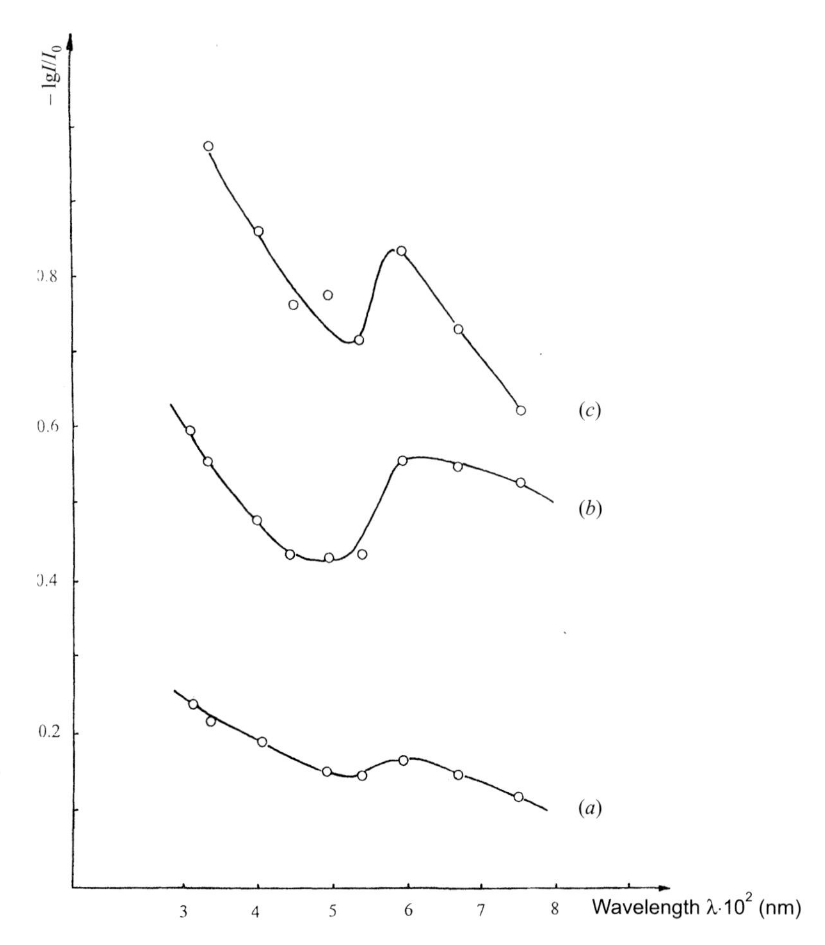

Fig. 4.2 The relation of optical density of milk and wavelength of optical radiation

spectrophotometer UR-20 was used for measurements and it was equipped with a CaF_2 cuvette. The collimated infrared radiation passed through the sample, the thickness of which was about several micrometers. The spectral range was 800–5,000 cm^{-1} (12.5–2 μm). The spectrum was recorded with the milk sample in place, and then with a reference sample with water substituting the sample, in order to eliminate the effect caused by the presence of water in milk samples, on the results of measurements.

4.4 RESULTS OF INFRARED SPECTROPHOTOMETRY OF MILK AND DAIRY PRODUCTS

Typical infrared (IR) absorption spectra of milk (3.9% fat), cream (10% fat) and sour cream (20% fat) are presented in Figs. 4.3–4.5; results of measurements of absorption bands are recorded in Table 4.1.

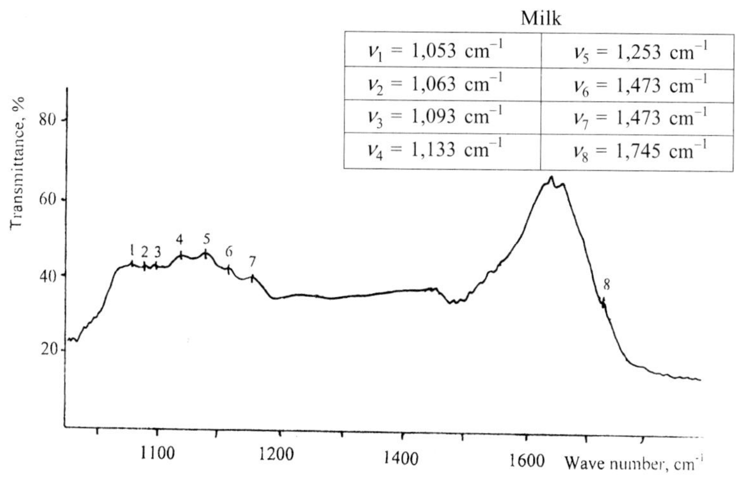

Fig. 4.3 Infrared absorption spectrum of milk (3.9% fat)

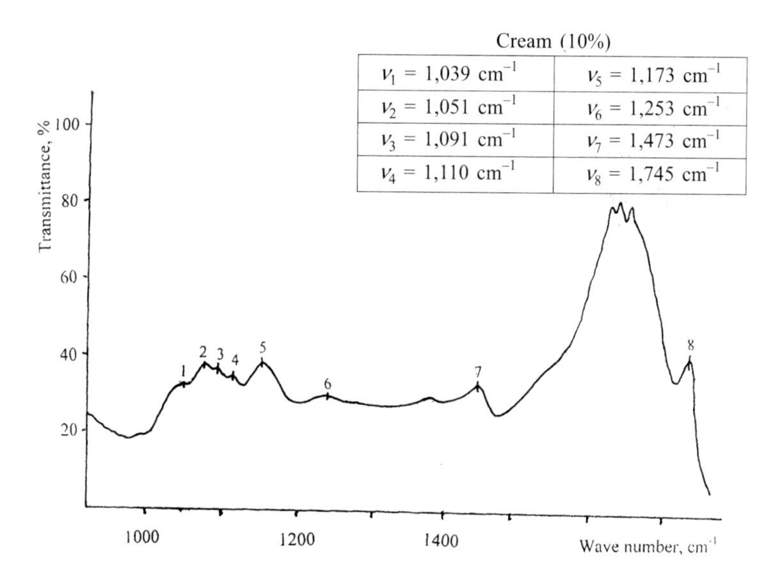

Fig. 4.4 Infrared absorption spectrum of cream (10% fat)

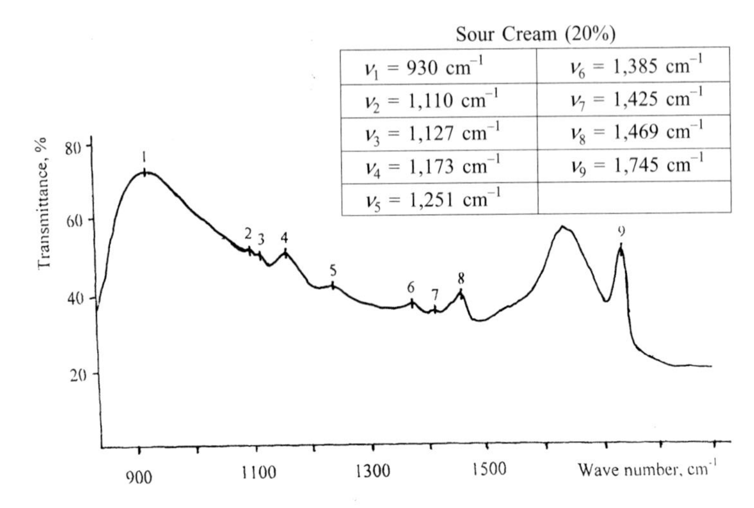

Fig. 4.5 Infrared absorption spectrum of sour cream (20% fat)

Table 4.1 Absorption bands of milk and dairy products [Posudin and Kostenko, 1994]

Absorption Band	*Milk*		*Cream*		*Sour Cream*	
	Wave Number (cm^{-1})	*Wavelength (μm)*	*Wave Number (cm^{-1})*	*Wavelength (μm)*	*Wave Number (cm^{-1})*	*Wavelength (μm)*
ν_1	1,053	9.50	1,039	9.62	930	10.75
ν_2	1,063	9.41	1,051	9.51	1,110	9.01
ν_3	1,093	9.15	1,091	9.17	1,127	8.87
ν_4	1,133	8.83	1,110	9.01	1,173	8.53
ν_5	1,173	8.53	1,173	8.53	1,251	7.99
ν_6	1,253	7.98	1,253	7.98	1,385	7.22
ν_7	1,473	6.79	1,473	6.79	1,425	7.02
ν_8	1,745	5.73	1,745	5.73	1,469	6.81
ν_9	-	-	-	-	1,745	5.73

Infrared spectra of boiled butter (Fig. 4.6) and aqueous solutions of lactose (Fig. 4.7) were also investigated; the values of absorption bands are given in Table 4.2. The position of absorption bands of aqueous solutions of lactose is stable in the range of solution concentration 0.1–0.8 mol/L.

It was shown that the position of absorption maxima ν_i of milk did not demonstrate any major spectral shifts at a given temperature, pH and chemical composition of the samples within the limits of measurement errors. The concentration C of fat is the most important factor that

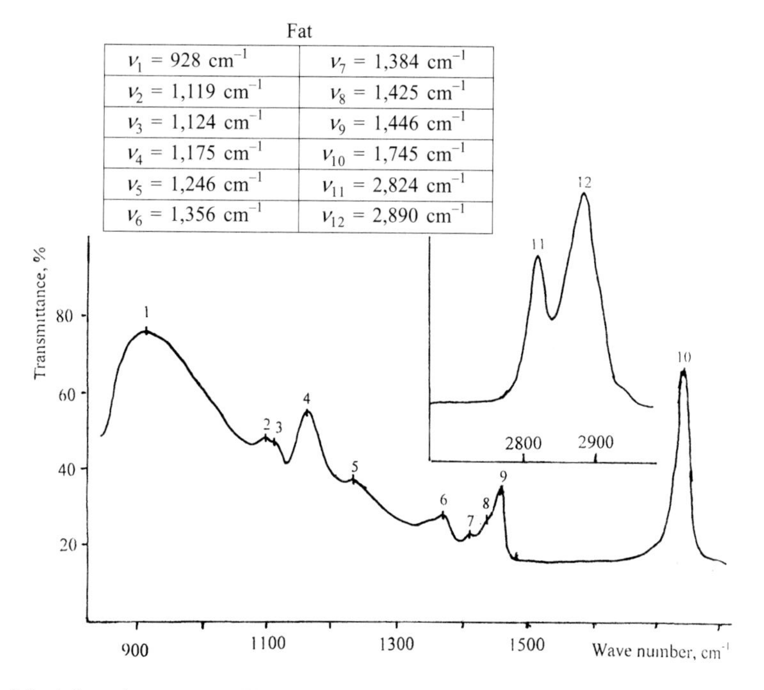

Fat

ν_1 = 928 cm^{-1}	ν_7 = 1,384 cm^{-1}
ν_2 = 1,119 cm^{-1}	ν_8 = 1,425 cm^{-1}
ν_3 = 1,124 cm^{-1}	ν_9 = 1,446 cm^{-1}
ν_4 = 1,175 cm^{-1}	ν_{10} = 1,745 cm^{-1}
ν_5 = 1,246 cm^{-1}	ν_{11} = 2,824 cm^{-1}
ν_6 = 1,356 cm^{-1}	ν_{12} = 2,890 cm^{-1}

Fig. 4.6 Infrared spectrum of boiled butter

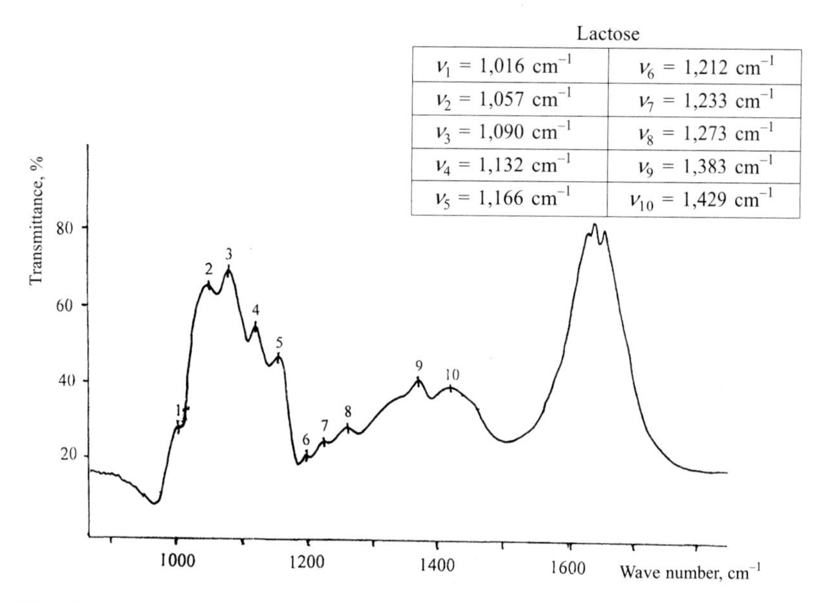

Lactose

ν_1 = 1,016 cm^{-1}	ν_6 = 1,212 cm^{-1}
ν_2 = 1,057 cm^{-1}	ν_7 = 1,233 cm^{-1}
ν_3 = 1,090 cm^{-1}	ν_8 = 1,273 cm^{-1}
ν_4 = 1,132 cm^{-1}	ν_9 = 1,383 cm^{-1}
ν_5 = 1,166 cm^{-1}	ν_{10} = 1,429 cm^{-1}

Fig. 4.7 Infrared spectrum of aqueous solutions of lactose

Table 4.2 Absorption bands of boiled butter and aqueous solutions of lactose [Posudin and Kostenko, 1994]

Absorption Band	Boiled Butter		Lactose Solution	
	Wave Number (cm^{-1})	Wavelength (μm)	Wave Number (cm^{-1})	Wavelength (μm)
ν_1	928	10.78	1,016	9.84
ν_2	1,119	8.94	1,057	9.46
ν_3	1,124	8.90	1,090	9.17
ν_4	1,175	8.51	1,132	8.83
ν_5	1,246	8.03	1,166	8.58
ν_6	1,356	7.37	1,212	8.25
ν_7	1,384	7.23	1,233	8.11
ν_8	1,425	7.02	1,273	7.86
ν_9	1,446	6.92	1,383	7.23
ν_{10}	1,745	5.73	–	–
ν_{11}	2,824	3.54	–	–
ν_{12}	2,890	3.46	–	–

provides effect on position of absorption bands (Fig. 4.8) and half-width of spectral line (Fig. 4.9) (the most isolated line $\nu_8 = 1{,}745\ cm^{-1}$ of milk has been chosen). The decreasing of fat concentration provokes the broadening of initial half-width (21.1 cm^{-1}) and its shift to short-wavelength region. It is interesting to note that the dependence of the ratio $D(1{,}473)/D(1{,}745)$ on the fat concentration demonstrates a linear character (Fig. 4.10); it is possible to use this ratio as a spectral index during measurement of fat in milk samples or for calibration procedure.

The process of turning sour causes the deformation of infrared absorption spectra and appearance of a broad maximum near 960 cm^{-1} (Fig. 4.11).

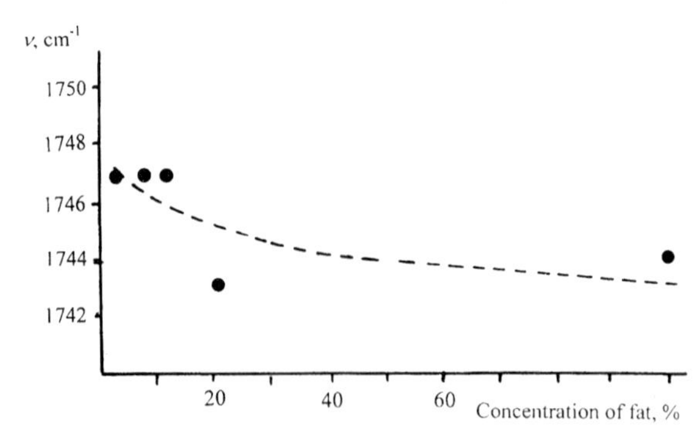

Fig. 4.8 Dependence of the position of absorption bands on the concentration of fat C (spectral line ν_8 = 1745 cm^{-1} of milk)

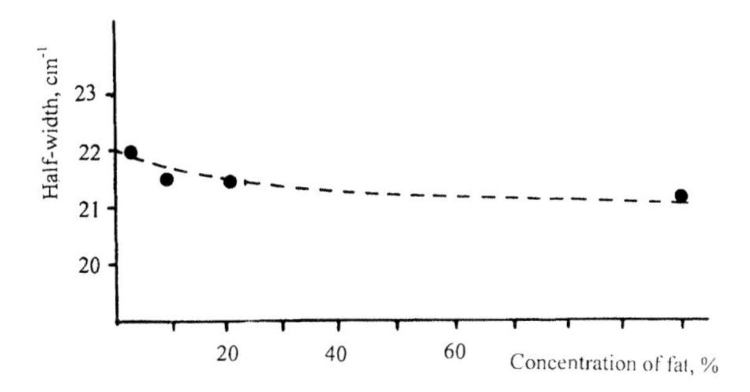

Fig. 4.9 Dependence of the half-width of spectral line (ν_8 = 1745 cm^{-1}) of milk on the concentration of fat *C*

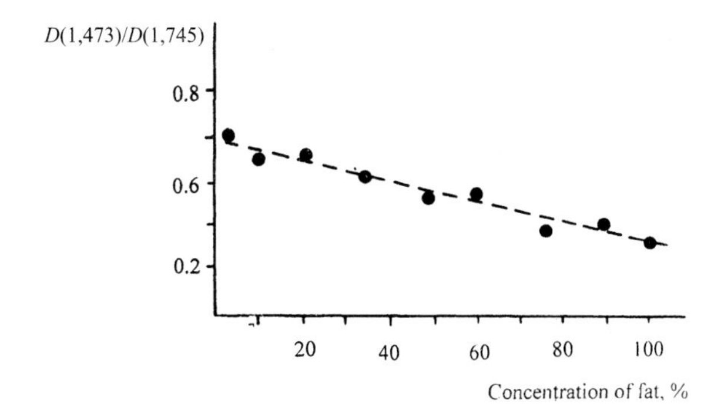

Fig. 4.10 Linear character of dependence of ratio *D*(1,473)/*D*(1,745) on the fat concentration

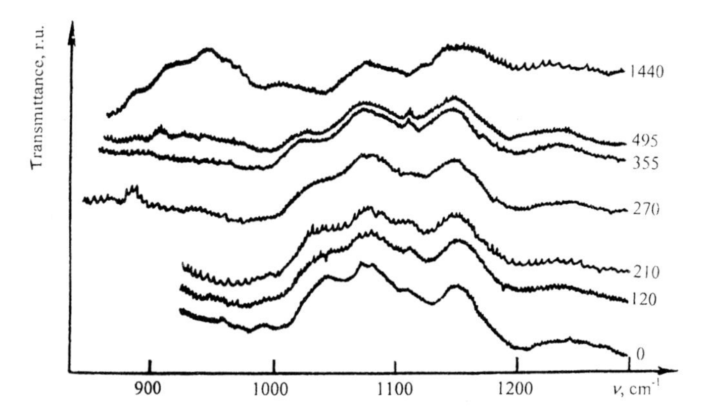

Fig. 4.11 The deformation of infrared absorption spectra and appearance of a broad maximum near 960 cm^{-1} due to the process of turning sour

4.5 INFRARED SPECTROPHOTOMETER FOR MILK ANALYSIS

An optical design of infrared spectrophotometer for milk analysis [Posudin and Kostenko, 1989, 1992, 1994; Martynenko et al., 1996; Posudin and Timoshenko, 1991; Posudin et al., 1997; Posudin, 2002, 2005] is illustrated in Fig. 4.12. The instrument consists of source *1* of infrared radiation, concave mirror *2*, diaphragm *3*, modulator *4* of radiation, lens *5*, cuvette *6* with a sample, interference filter *7*, detector *8*, power supply *9* and readout system 10.

Source of Infrared Radiation. A nichrome coiled filament that was used as the main component of the source of infrared radiation is characterized by the following parameters: voltage 6–8 V; electric current 2.5–3.0 A; power 15–24 W. Maximum of spectral emission is near 3–6 μm. Diameter of filament 0.4 mm, and the length 23 mm; diameter and length of the coil 5 and 12 mm correspondingly. The source was located in the focus of a concave aluminium mirror to provide maximal reflection and radiation density.

A Sample Cuvette. The windows of cuvette (thickness 300 μm) were performed with germanium; the total transmission of such a system was 70%. The air gap between the windows was equal 150 μm; such a value made it possible to avoid possible interference effects with a narrow gap and to keep the sensitivity of readout system.

Interference Filters. It was necessary to use the interference filters which were tuned at the maximum and at the base of absorption band of each component of milk (Fig. 4.13). The main parameters of interference filters are presented in Table 4.3.

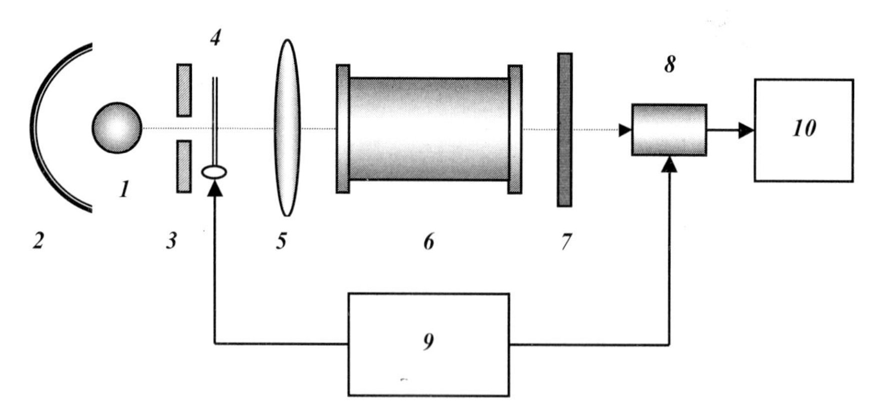

Fig. 4.12 An optical design of infrared spectrophotometer for milk analysis [Posudin and Kostenko, 1994]. 1 – source of infrared radiation, 2 – concave mirror, 3 – diaphragm, 4 – modulator of radiation, 5 – lens, 6 – cuvette with a sample, 7 – interference filter, 8 – detector, 9 – power supply; 10 – readout system

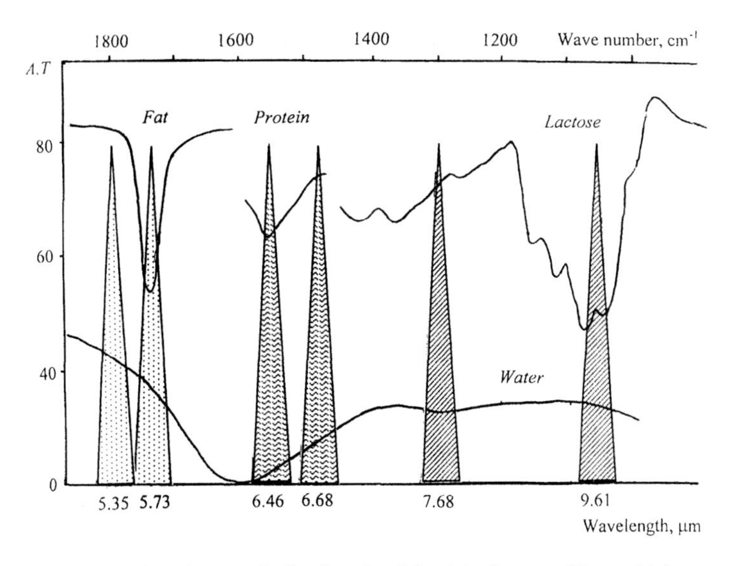

Fig. 4.13 The position of transmission bands of the interference filters which were tuned at the maximum and at the base of absorption band of each component of milk

Table 4.3 Parameters of interference filters which were used in infrared spectrophotometer [Posudin and Kostenko, 1994]

Position Relative to Absorption Band of Milk Component	λ *(μm)*	*T (%)*	$\Delta_{0.5}$ *(μm)*	$\Delta_{0.1}$ *(μm)*
Basis of absorption band of water	4.50 + 0.03	43	0.11	0.42
Maximum of absorption band of water	4.42 – 0.05	43 – 44	0.10	0.35
Basis of absorption band of fat	5.35 – 0.02	44 – 45	0.13	0.38
Maximum of absorption band of fat	5.70 + 0.03	42	0.17	0.46
Basis of absorption band of protein	6.68 – 0.02	40 – 43	0.22	0.56
Maximum of absorption band of protein	6.46 – 0.04	40	0.18	0.57
Basis of absorption band of lactose	7.68 – 0.03	50	0.23	0.54
Maximum of absorption band of lactose	9.61 + 0.01	42	0.26	0.65

If the direction of radiation flow is deviating from normal to the surface of the interference filter, the spectral shift of absorption band and broadening of this band can take place. The value of the shift depends on the angle of incidence and refractive index of filter material. The dependence of refractive index of germanium on the wavelength of radiation can be expressed by the following equation [Voronkova et al., 1965]:

$$N = A + BL + CL^2 + D\lambda^2 + E\lambda^4 \qquad \text{(Eqn. 4.1)}$$

where A = 3.99931; B = 0.391707; C = 0.163492; D = 0.000006; E = 0.00000053; $L = (\lambda^2 - 0.028)^{-1}$.

Thus, refractive index according to this equation is changing from 4.108 (λ = 2 µm) to 4.005 (λ = 8 µm).

The values of spectral shift $\Delta\lambda_\theta/\lambda_0$ can be calculated due the following expression:

$$\Delta\lambda_\theta/\lambda_0 = [(n^2 - \sin^2\theta)/n]^{1/2} \quad \text{(Eqn. 4.2)}$$

where λ_0–wavelength of incident radiation.

These values are presented in Table 4.4.

Table 4.4 Dependence of spectral shift of absorption band on the deviation of radiation flow direction from normal to the surface of interference filter [Posudin and Kostenko, 1994]

$\theta°$	λ = 2 µm (n = 4.108)	λ = 2 µm (n = 4.024)
0	1	1
5	0.99977	0.99976
10	0.99910	0.99907
15	0.99800	0.99790
20	0.99650	0.99640
25	0.99470	0.99450

Readout System. Infrared radiation that was transmitted by the sample entered the pyrometric element PIP with the following characteristics: spectral range 0.5–15 µm; coefficient of transformation $2{\cdot}10^3$ V/W; sensitivity threshold 10^{-8}–10^{-9} W; area of sensitive surface 2 × 2 mm^2.

Procedure of Measurement. The concentration of milk component is determined as follows:

$$C_i = K\ln(1 - \beta_i^{max} + \beta_i^0) \quad \text{(Eqn. 4.3)}$$

where K – coefficient of proportionality; β_i^{max} and β_i^0 – values of absorption at the maximum, and at the base of absorption band of *i*-component of milk.

The effect of water absorption is taken into account by the following formula:

$$\beta_i^{\lambda_i}(\text{sample}){\cdot}100\% = [I_i^{\lambda_i}(H_2O) - I_i^{\lambda_i}(\text{sample})]/I_i^{\lambda_i}(H_2O) \quad \text{(Eqn. 4.4)}$$

where $I_i^{\lambda_i}$ (sample) and $I_i^{\lambda_i}$ (H_2O) – intensity of absorption band for milk sample and water at the wavelength λ_i.

4.6 NEAR-INFRARED SPECTROSCOPY OF MILK AND MILK COMPONENTS

The main components of milk (fat, casein, lactose) have specific absorption bands in NIR part of spectrum (Table 4.5).

Table 4.5 Specific absorption bands of milk components in NIR part of spectrum [Fuhrman, 1987]

Absorption Maximum (nm)	*Component*	*Absorption Maximum (nm)*	*Component*
2,340	Fat, casein, lactose	2,100	Casein, lactose, water
2,310	Fat, casein	1,340	Water
2,270	Fat, casein	1,730	Fat, casein
2,790	Casein	1,720	Fat, casein
1,980	Casein, water	1,450	Casein, lactose, water
1,820	Casein, lactose, fat	1,680	Casein, water
1,780	Fat, casein		

4.7 INSTRUMENTATION

Near-infrared spectrum of milk was investigated with infrared analyzer 4250 "Pacific Scientific". This device has three spectral regions at 1,620–1,800 nm; 1,890–2,115 nm; 2,050–2,320 nm. The following grading equation was used for elaboration of the results of measurements:

$$Y = B_0 + B_1 \cdot OP(\lambda_1) + B_2 \cdot OP(\lambda_2) + \cdots + B_N \cdot OP(\lambda_N) \qquad \text{(Eqn. 4.5)}$$

where Y – result of infrared analysis; B_0, B_1, ..., B_N – coefficients of grading equation; $OP(\lambda_i)$ – optical parameters of spectrum at a given wavelength. Optical density D was used as an optical parameter:

$$D = \lg[1/R(\lambda)] \qquad \text{(Eqn. 4.6)}$$

where $R(\lambda)$ is diffused reflectance at wavelength λ.

In addition, the analyzer provides the determination of first D_1 and second D_2 derivatives:

$$D_1 = \lg[1/R(\lambda - d\lambda)] - \lg[1/R(\lambda + d\lambda)] \qquad \text{(Eqn. 4.7)}$$

$$D_2 = \lg[1/R(\lambda - d\lambda)] - 2\lg[1/R(\lambda)] + \lg[1/R(\lambda + d\lambda)] \qquad \text{(Eqn. 4.8)}$$

where $d\lambda$ is a step of derivative.

4.8 RESULTS OF NEAR-INFRARED SPECTROSCOPY OF MILK

The main objective of this investigation was to determine the principal spectral characteristics of milk samples with different contents of components, to estimate level of correlation between these components, and to compare the results of determination of milk components which were obtained due to NIR method and traditional chemical methods.

The following coefficient of correlation R was used for comparison of the results of measurements by NIR (Y_{NIR}) and chemical (Y_{chem}) methods:

$$R = \frac{N\sum(Y_{chem} \cdot Y_{NIR}) - \sum(Y_{chem} \cdot Y_{NIR})}{[N\sum Y_{chem}^2 - (\sum Y_{chem})^2]^{1/2} \times [N\sum Y_{NIR}^2 - (\sum Y_{NIR})^2]^{1/2}}$$

(Eqn. 4.9)

The results of chemical and NIR analysis of 50 samples of milk are given in Table 4.6 and the results of calculation of coefficients of grading equation (Eqn. 4.5) are presented in Table 4.7.

The correlation dependence between results of chemical and NIR analysis of milk components are presented in Fig. 4.14 for fat, Fig. 4.15 for protein, Fig. 4.16 for non-fat solids and Fig. 4.17 for total solids.

The values of coefficients of correlation which were obtained during analysis of 50 samples of milk are given in Table 4.8.

Near-infrared spectra of milk samples with different contents of fat (1.82%; 4.27%; 4.48%; 4.73%; 7.69%) are presented in Fig. 4.18. It is clear that the amplitude of these spectra depends strongly on the fat content. It was also interesting to compare the NIR spectra of the samples that had the same fat (Fig. 4.19) and protein (Fig. 4.20) concentration according to the chemical testing: the results demonstrate some divergence which are caused by inaccuracy of the chemical method. The derivative spectra are rather informative – they make it possible to find extreme points of the reflectance spectra (samples N 1, 5 and 7) (Fig. 4.21).

The next step of investigation was the clearing up of the level of correlation between separate components of milk which were determined by chemical and NIR methods. The correlation dependence between the main components of milk (fat, protein, total solids, and non-fat solids) are presented in Figs. 4.22–4.27. The highest level of correlation was indicated between the content of fat and total solids that was determined by NIR method (Fig. 4.27). The values of coefficients of correlation are given in Table 4.9.

Table 4.6 Results of chemical and infrared analysis

Sample number	*Fat*	*Protein*	*Dry residue*	*Dry substance*	*Sample number*	*Fat*	*Protein*	*Dry residue*	*Dry substance*
1	8.40	2.90	8.20	16.60	1	7.69	3.44	8.76	16.68
2	4.40	3.70	8.60	13.00	2	4.73	3.02	8.25	12.92
3	4.10	2.90	8.30	12.40	3	4.27	2.67	7.98	12.04
4	4.50	2.50	7.90	12.40	4	4.48	2.65	8.00	12.32
5	1.40	3.80	9.10	10.50	5	1.82	3.30	8.46	9.91
6	3.80	2.70	8.00	11.80	6	3.86	2.76	8.16	12.14
7	4.40	3.50	8.70	13.10	7	4.45	2.61	8.80	12.25
8	5.40	3.20	8.70	14.10	8	5.36	2.95	8.39	13.44
9	4.00	2.70	8.00	12.00	9	4.14	2.66	8.17	12.24
10	5.00	3.00	8.20	13.20	10	5.05	2.88	8.26	13.29
11	2.10	2.30	7.50	9.60	11	2.03	2.99	7.98	10.07
12	2.65	2.80	8.00	10.65	12	2.44	2.80	7.99	10.40
13	3.90	2.40	7.50	11.40	13	3.74	2.58	7.97	11.75
14	3.90	3.20	8.60	12.50	14	3.99	2.63	8.19	11.87
15	3.50	2.70	7.50	11.00	15	3.60	2.71	8.23	11.53
16	4.50	2.70	7.60	11.10	16	4.43	2.88	7.93	12.35
17	5.70	3.00	8.30	14.00	17	4.91	2.87	8.14	13.34
18	4.10	3.40	8.70	12.80	18	4.14	2.69	8.10	11.83
19	3.30	2.80	7.90	11.20	19	3.70	3.37	8.38	11.18
20	3.80	3.00	8.30	12.10	20	3.86	3.01	8.33	12.40
21	5.00	3.20	8.70	13.70	21	5.20	2.75	8.32	13.56
22	3.50	2.60	8.10	11.60	22	4.33	2.61	7.81	11.67
23	4.00	2.80	8.10	12.10	23	4.14	2.75	8.36	12.31
24	3.10	2.80	8.00	11.10	24	3.09	2.78	8.18	11.40
25	4.00	2.00	7.30	11.30	25	3.94	2.47	7.84	11.57
26	3.30	2.60	7.50	10.80	26	2.85	2.59	7.48	10.76
27	3.70	2.45	7.40	11.10	27	3.52	2.61	7.79	11.32
28	3.20	2.10	7.20	10.40	28	3.03	2.50	7.52	10.46
29	4.70	3.10	8.40	13.10	29	4.69	2.87	7.89	12.93
30	2.70	2.70	7.90	10.60	30	2.72	2.78	7.76	10.63
31	4.00	2.90	8.30	12.30	31	4.06	2.70	7.88	12.01
32	3.90	3.00	8.20	12.10	32	3.80	2.92	7.83	12.18
33	4.10	2.25	7.30	11.40	33	4.85	2.57	7.53	12.46
34	3.40	2.50	7.60	11.00	34	3.36	2.95	8.22	11.76
35	3.30	2.50	7.60	10.90	35	3.26	2.50	7.66	11.30
36	3.90	2.90	8.30	12.20	36	3.90	2.96	8.03	12.39
37	3.20	1.20	7.50	10.70	37	3.15	2.51	7.62	10.94
38	3.90	2.50	8.40	12.30	38	3.80	2.57	7.79	11.38
39	4.60	2.40	7.60	12.20	39	4.59	2.43	7.66	11.96

(*Table Contd.*)

(*Table Contd.*)

40	4.90	3.10	8.00	12.90	40	4.89	2.62	7.76	12.58
41	3.90	2.60	7.70	11.60	41	3.85	2.52	7.68	11.37
42	3.20	2.65	7.80	11.00	42	3.16	2.64	7.60	10.79
43	4.60	2.90	8.20	12.80	43	4.65	2.94	7.78	12.55
44	2.80	2.20	7.30	10.10	44	2.52	2.52	1.58	10.18
45	4.60	2.35	7.40	12.00	45	4.36	2.65	7.75	12.12
46	3.50	2.90	7.70	11.20	46	3.20	2.70	7.73	11.51
47	5.00	2.80	7.30	12.30	47	5.62	2.72	7.52	13.33
48	4.10	3.20	8.00	12.10	48	3.96	2.77	7.77	12.06
49	4.70	2.60	7.60	12.30	49	4.50	2.58	7.64	11.95
50	4.00	2.70	7.50	11.50	50	3.96	2.71	7.77	11.78

Table 4.7 Results of calculation of coefficients of grading equation (Eqn. 4.5) [Posudin and Kostenko, 1994]

Coefficient of Regression Equation	*F-criterion for Given Equation*	*Scale Division*	*Wave Number, cm^{-1}*
	Protein		
B(0) = 8.719			
B(1) = 64.680	14.02	285	2,197
B(2) = 230.190	14.79	93	1,792
B(3) = –69.974	19.95	35	1,705
B(4) = –73.980	62.69	264	2,131
	Fat		
B(0) = 6.918			
B(1) = 316.151	44.84	88	1,789
B(2) = –108.606	506.08	24	1,679
	Non-fat Solids		
B(0) = 14.539			
B(1) = –36.935	58.48	186	2,062
B(2) = –82.348	77.57	24	1,679
	Total Solids		
B(0) = 21.224			
B(1) = –81.700	35.03	271	2,154
B(2) = 135.838	11.23	293	2,219
B(3) = 243.213	21.40	350	2,315
B(4) = –172.508	66.10	28	1,688

NIR method provides the analysis of milk samples which contain a high proportion of water and demonstrate high level of opacity. The NIR method is rapid, non-destructive and offers a high level of accuracy in comparison with traditional chemical methods.

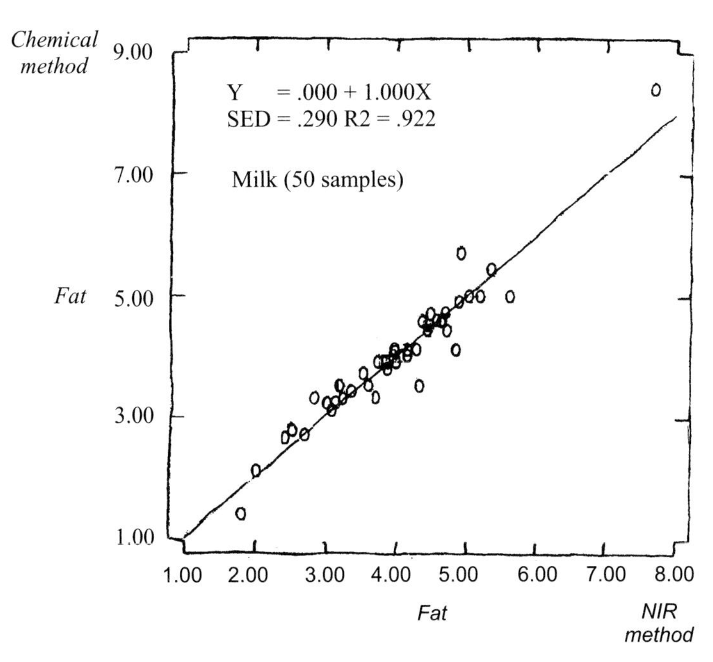

Fig. 4.14 The correlation dependence between results of chemical and NIR analysis of fat in milk

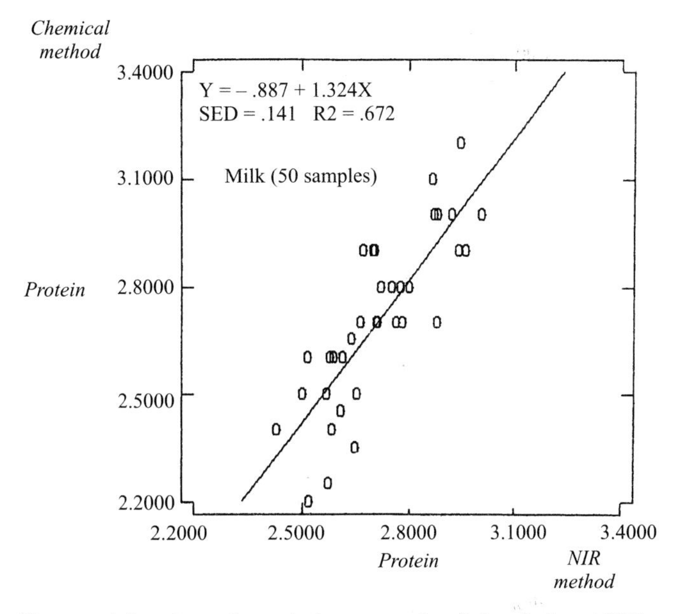

Fig. 4.15 The correlation dependence between results of chemical and NIR analysis of protein in milk

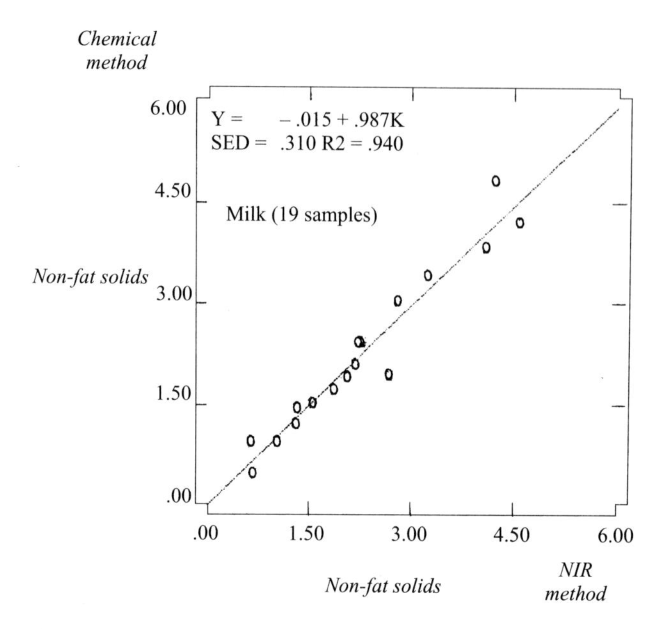

Fig. 4.16 The correlation dependence between results of chemical and NIR analysis of non-fat solids in milk

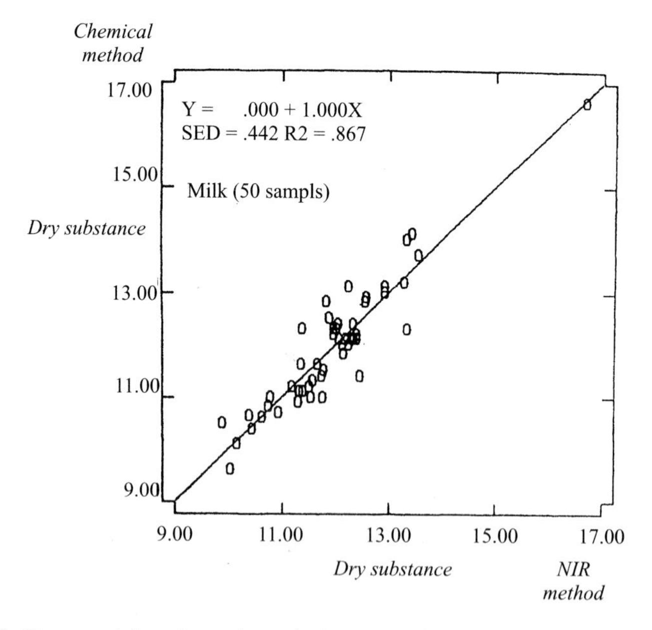

Fig. 4.17 The correlation dependence between results of chemical and NIR analysis of total solids in milk

Table 4.8 Values of coefficients of correlation dependence between results of chemical and NIR analysis of milk components [Posudin and Kostenko, 1994]

Type of Analysis	*Component*	*Mean Content (%)*	*Deviation*	*SEP*	*SEP(C)*	*R*	R_2
Chemical	Fat	3.99	1.028	0.284	0.287	0.960	0.922
NIR	Fat	3.99	0.987				
Chemical	Protein	2.75	0.439	0.375	0.379	0.504	0.254
NIR	Protein	2.75	0.221				
Chemical	Non-fat Solids	7.95	0.465	0.357	0.361	0.630	0.397
NIR	Non-fat Solids	7.95	0.293				
Chemical	Total Solids	11.94	1.198	0.433	0.437	0.931	0.867
NIR	Total Solids	11.94	1.115				

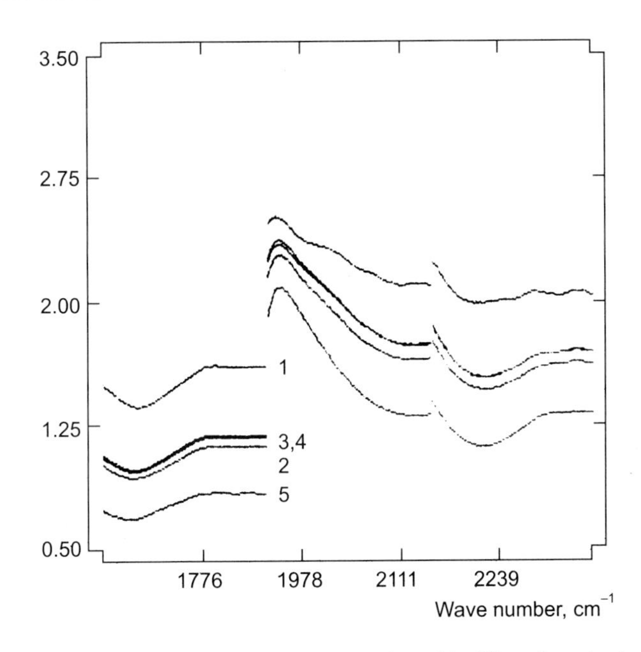

Fig. 4.18 The near-infrared spectra of milk samples with different contents of fat (1 – 1.82%; 2 – 4.27%; 3 – 4.48%; 4 – 4.73%; 5 – 7.69%)

4.9 FLUORESCENCE SPECTROSCOPY OF MILK AND MILK COMPONENTS

Milk contains a number of *fluorophors* – compounds which can produce fluorescence. The wavelengths of excitation and emission of fluorescence of the main milk fluorophores at room temperature are presented in

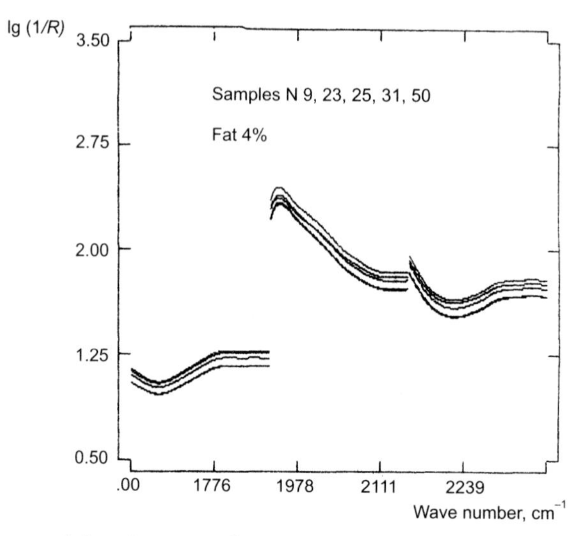

Fig. 4.19 The near-infrared spectra of the samples that had the same fat concentration according to the chemical testing

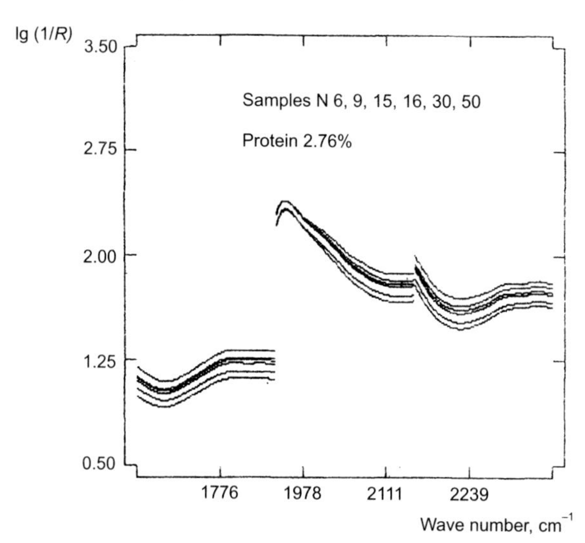

Fig. 4.20 The near-infrared spectra of the samples that had the same protein concentration according to the chemical testing

Table 4.10. The fluorescence properties of amino acids and proteins depend on the solvent, the substituent groups, pH of the medium and the temperature. The process of oxidation and photodegradation cause

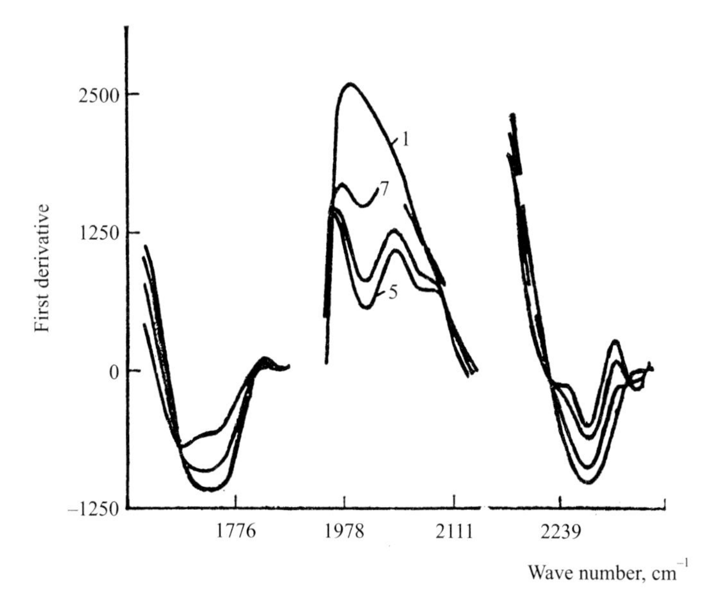

Fig. 4.21 The derivative spectra of milk (samples N 1, 5 and 7)

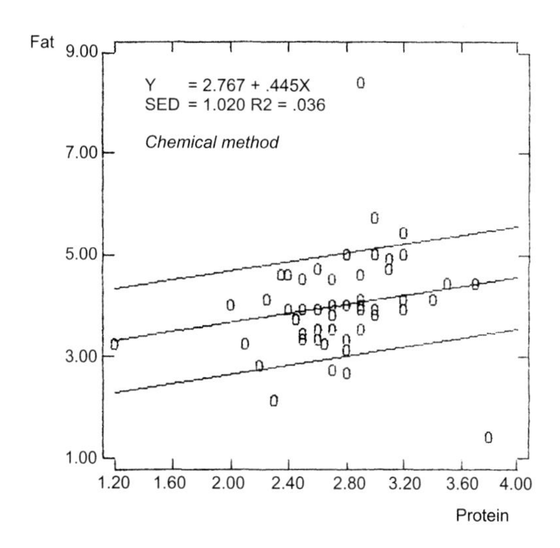

Fig. 4.22 The correlation between fat and protein content of milk which was determined by chemical method

the change in fluorescence intensity and spectral shifts in fatty acids. The fluorescence of soluble vitamins (B_1, B_2, B_{12}, *PP*, *H*) depends on the temperature, solvent and pH of the medium.

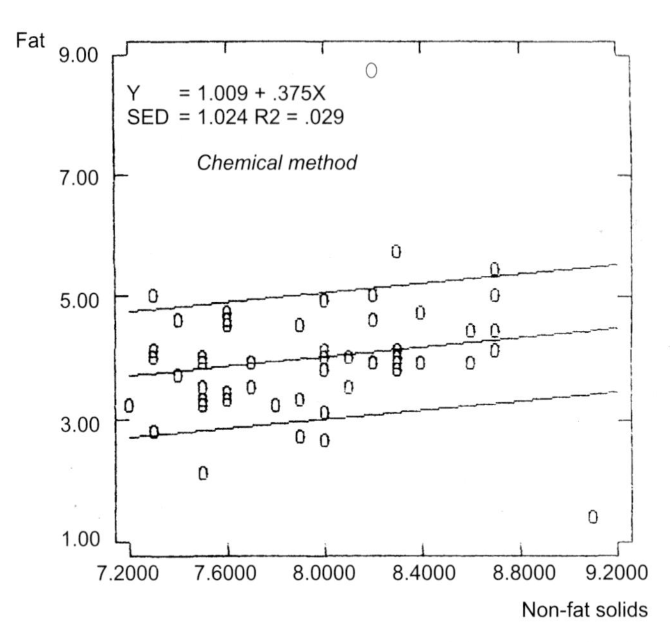

Fig. 4.23 The correlation between fat and non-fat solids content of milk which was determined by chemical method

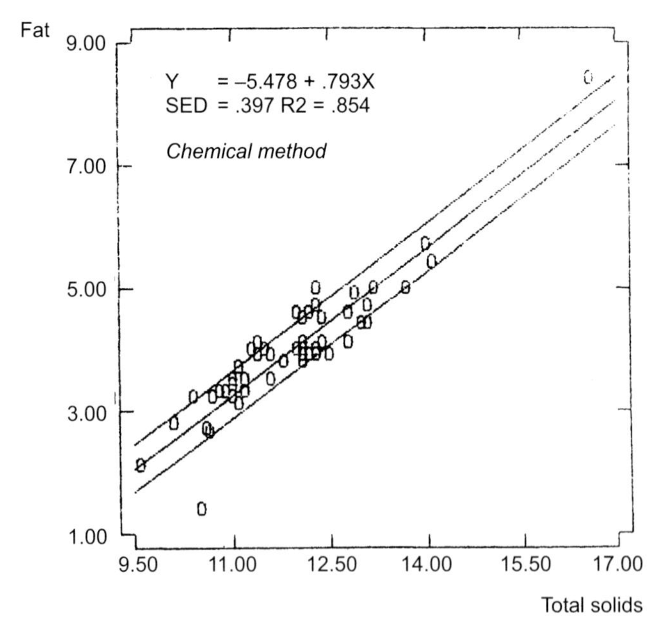

Fig. 4.24 The correlation between fat and total solids content of milk which was determined by chemical method

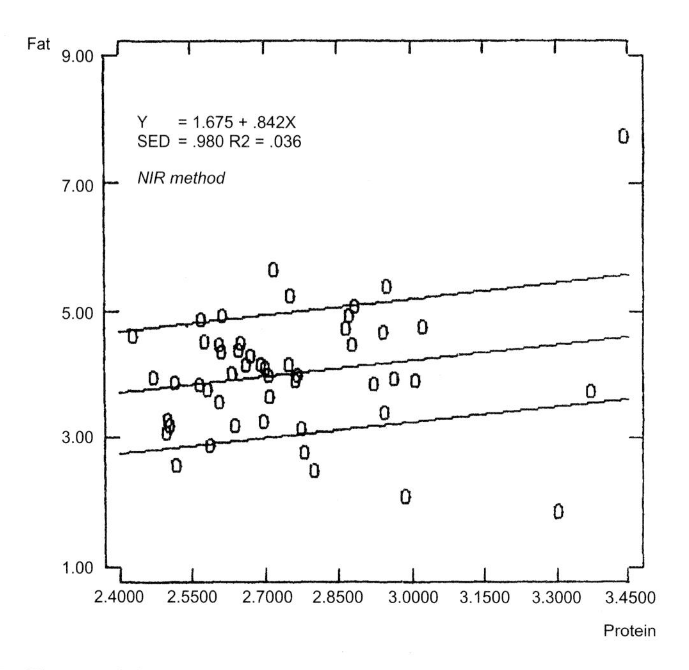

Fig. 4.25 The correlation between fat and protein content of milk which was determined by NIR method

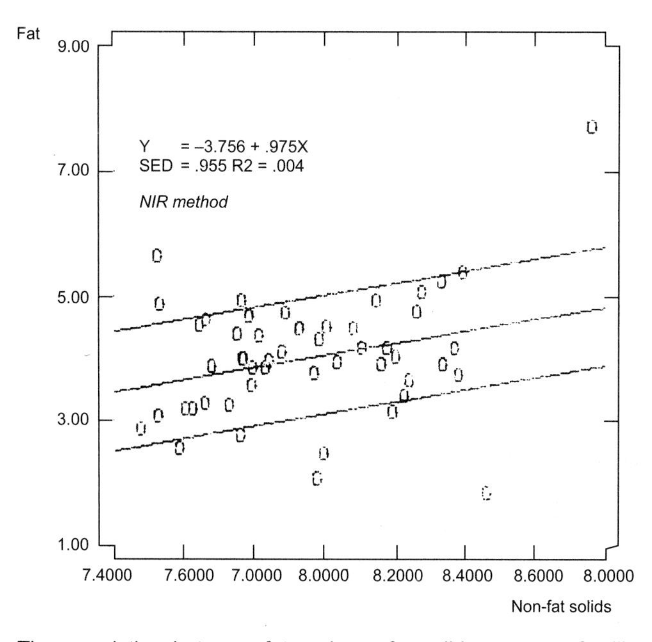

Fig. 4.26 The correlation between fat and non-fat solids content of milk which was determined by NIR method

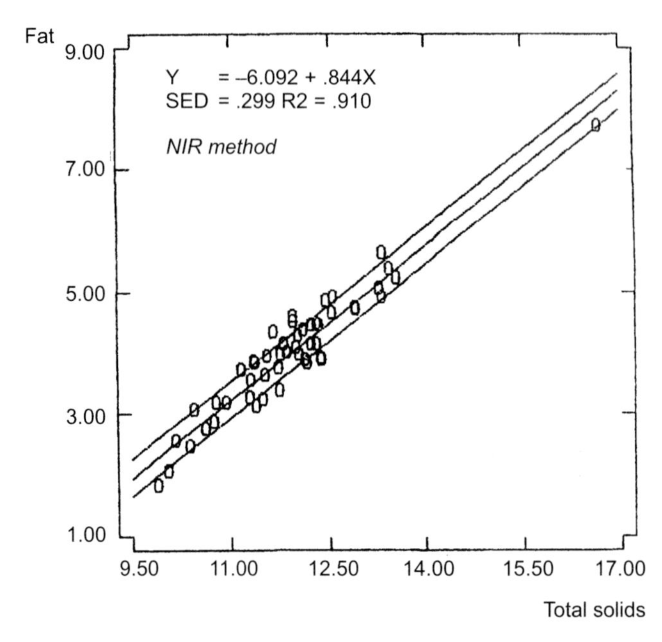

Fig. 4.27 The correlation between fat and total solids content of milk which was determined by NIR method

Table 4.9 Values of coefficients of correlation between separate components of milk which were determined by chemical and NIR methods [Posudin and Kostenko, 1994]

Component	*Fat*	*Protein*	*Non-fat Solids*	*Total Solids*
Fat	-	0.036	0.084	0.910
Protein	0.036	-	0.519	0.123
Non-fat Solids	0.084	0.519	-	0.200
Total Solids	0.910	0.123	0.200	-

4.10 INSTRUMENTATION

Fluorescence properties of milk and its components were measured by spectrophotometer SDL-2 at room temperature in spectral range 280-700 nm. Xenon lamp DKeSh-150 was used as a source of fluorescence excitation. The fluorescence emission radiation was registered at the angle 90° to the direction of excitation radiation. The photomultiplier FEU-100 was used for the detection of fluorescence emission. The absolute error of the determination of spectral line position was ±2 nm

Table 4.10 Dependence of fluorescence emission of milk components on the excitation wavelength

Wavelength of Excitation (nm)	*Maximum of Emission (nm)*	*Component (Responsible for Fluorescence)*
270-290	340	Amino acids, proteins
	370	Folic acid
	340	Vitamin *E*
300-310	340	Proteins (casein)
	520	Riboflavin
315-320	398	Vitamin B_{12}
	520	Riboflavin
	400	Fat acids
320-340	400	Vitamin B_6
	400	Vitamin B_{12}
	510	Vitamin *A*
	520	Riboflavin
340-370	424	Linetol
	450	Thiochrome (B_1)
	450	Folic acid
	460	Vitamin *C*
	460	Vitamin B_3
	471	Vitamin B_6
	480	Vitamin *D*
	520	Riboflavin
400	520	Riboflavin

and of half-width of spectral line was not more than ±5 nm. The relative error of determination of the intensity of spectral lines did not exceed ±2%.

4.11 RESULTS OF FLUORESCENCE SPECTROSCOPY OF MILK

4.11.1 Whole Milk

The problem of determination of proteins and fat in milk can be resolved by the selection of wavelength of fluorescence excitation. The typical spectra of fluorescence emission of milk for different wavelengths of excitation (Figs. 4.28 *a* and *b*) demonstrate that the selection of proteins or fat can be achieved by the choice of excitation at 280 or 325 nm correspondingly. The emission of fluorescence can be analyzed at 330–340 nm for proteins and at 400–410 nm for fat.

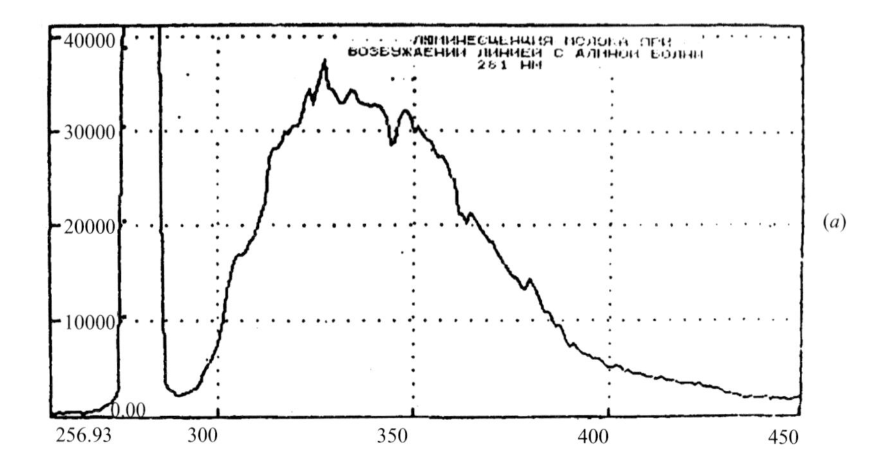

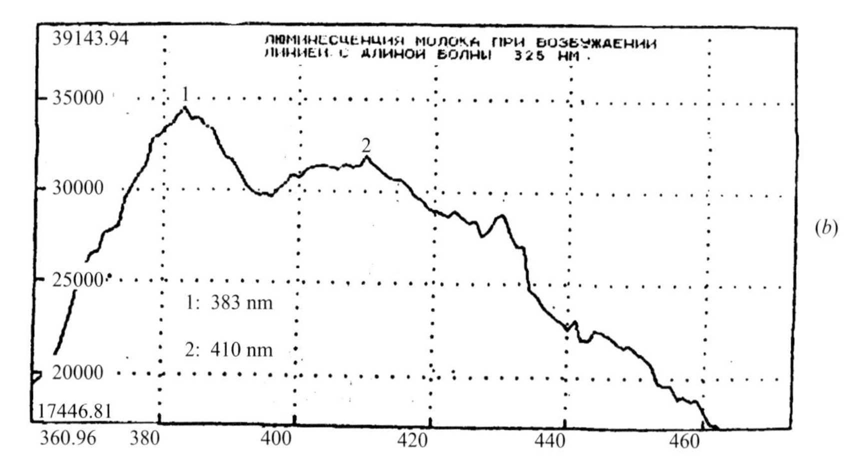

Fig. 4.28 The typical spectrum of fluorescence emission of milk: *a* – wavelength of excitation 281 nm; *b* – wavelength of excitation 325 nm

4.11.2 Proteins

Aromatic Amino Acids. The main amino acids of milk such as tryptophan, tyrosine, and phenylalanine, demonstrate fluorescence properties. Fluorescence emission spectrum of tryptophan (Fig. 4.29) was measured by us; it is characterized by maxima at 289 nm, 308 nm; 334 nm and 360 nm (excitation wavelength is 280 nm). Tyrosine in aqueous solutions demonstrates at room temperature fluorescence band of half-width of about 34 nm at 303–304 nm (excitation wavelength is 275 nm);

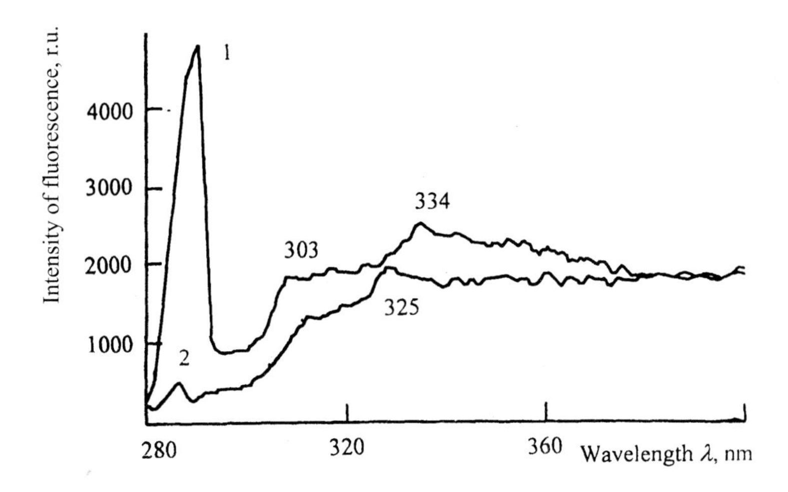

Fig. 4.29 Fluorescence emission spectrum of tryptophan (excitation wavelength is 280 nm). Wavelength of excitation: 1 – 280 nm; 2 – 313 nm

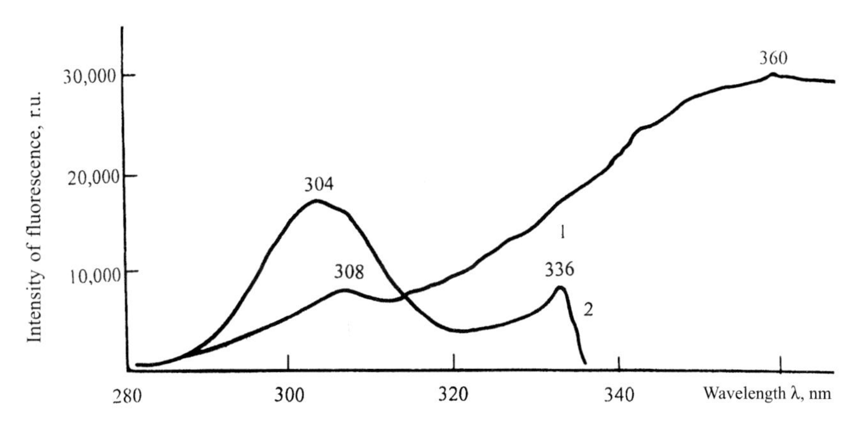

Fig. 4.30 Fluorescence excitation spectra of casein (emission wavelength: 1 – 350 nm; 2 – 400 nm)

phenylalanine has fluorescence emission maxima at 275 nm, 282 nm and 289 nm [Brusilovsky and Waynberg, 1990].

Casein. This component constitutes about 83% of all protein components of milk. Excitation fluorescence spectra of casein are given in Fig. 4.30; maxima of excitation spectrum are located at 304 nm and 336 nm (emission wavelength is 350 nm); fluorescence emission spectra of casein (Fig. 4.31) have maxima at 334–341 nm (excitation wavelengths are 290 nm, 295 nm; 300 nm; and 305 nm).

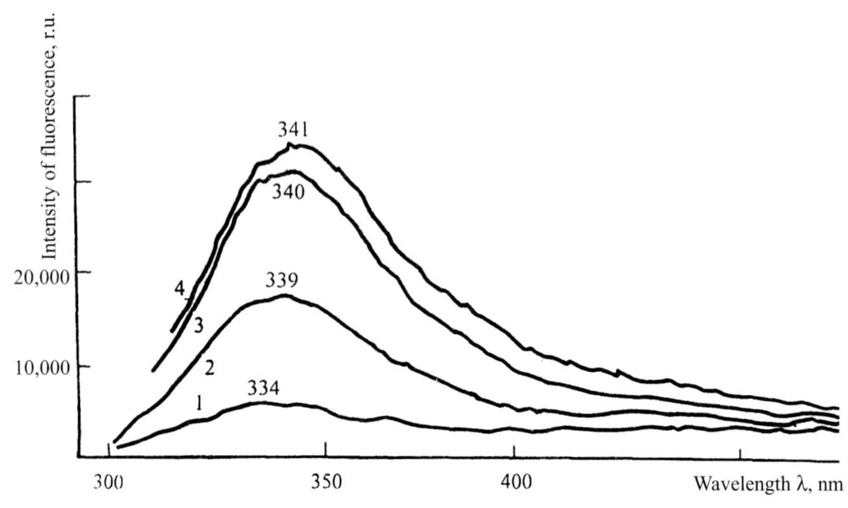

Fig. 4.31 Fluorescence emission spectra of casein (excitation wavelengths: 1 – 290 nm, 2 – 295 nm; 3 – 300 nm; 4 – 305 nm)

4.11.3 Fat Acids

Milk fat consists of polyunsaturated acids - linoleic acid with two unsaturated bonds, linolenic acid with three double bonds, and arachidonic acid with four double bonds. We have used linetol (mixture of these three acids) as a model system in our investigations. Fluorescence excitation spectrum of linetol is characterized by maximum at 362 nm with half-width of about 30 nm (Fig. 4.32); fluorescence emission spectrum has a maximum at 424 nm (Fig. 4.33) with half-width of about 100 nm. A strong effect of oxidation on the intensity of fluorescence of linetol - 7% decreasing during 10 minutes, was observed in our experiments.

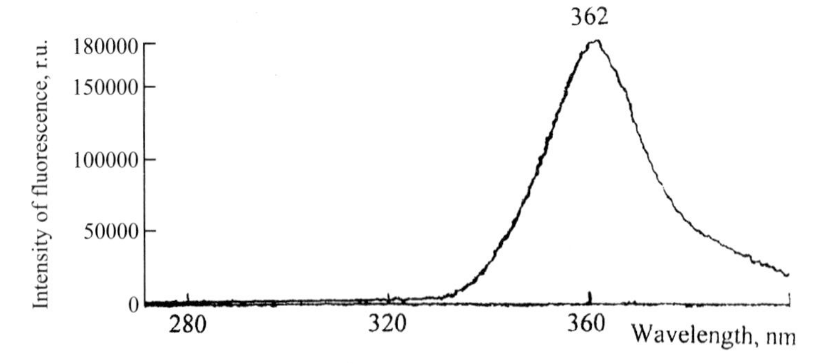

Fig. 4.32 Fluorescence excitation spectrum of linetol (emission wavelength is 425 nm)

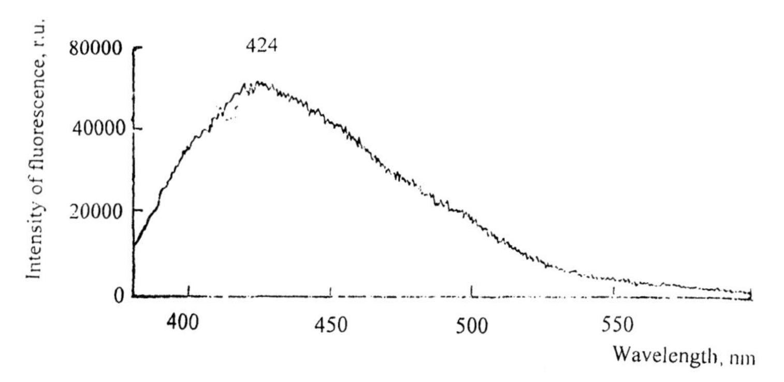

Fig. 4.33 Fluorescence emission spectrum of linetol (excitation wavelength is 365 nm)

4.11.4 VITAMINS

The main results of our own investigations of fluorescence properties of milk vitamins are presented in this section. In addition, the literature references are included to summarize fluorescence data for milk vitamins at room temperature.

Vitamin C (ascorbic acid) in aqueous solution demonstrates fluorescence excitation maximum at 369 nm with half-width of about 70 nm (Fig. 4.34) and fluorescence emission maximum at 460 nm with half-width of about 100 nm (Fig. 4.35).

Vitamin B_2 (riboflavine) is the most intense fluorophore of milk. Fluorescence excitation spectrum demonstrates the maxima at 370, 415

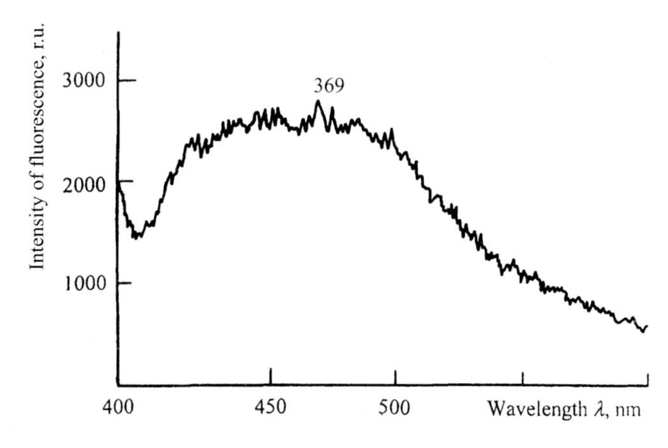

Fig. 4.34 Fluorescence excitation spectrum of vitamin C (ascorbic acid) in aqueous solution

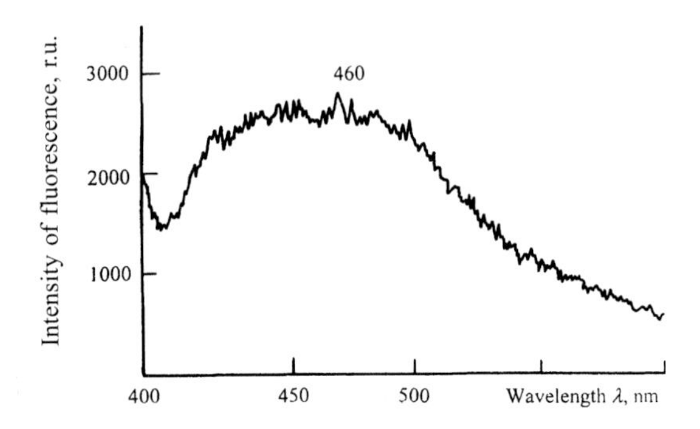

Fig. 4.35 Fluorescence emission spectrum of vitamin *C* (ascorbic acid) in aqueous solution

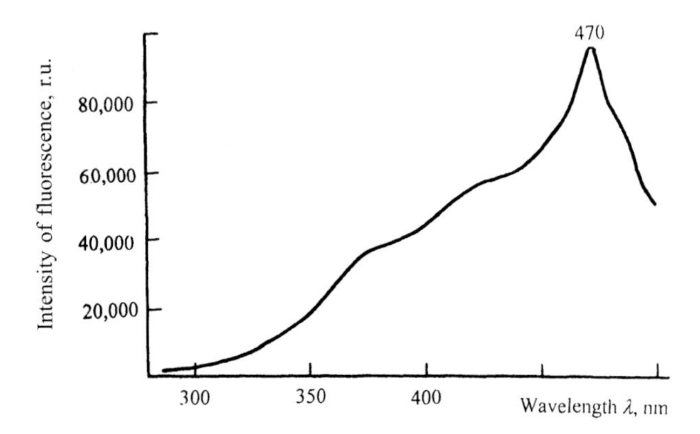

Fig. 4.36 Fluorescence excitation spectrum of vitamin B_2 (riboflavine) (emission wavelength is 520 nm)

and 470 nm (emission wavelength is 520 nm) and 469 and 538 nm (emission wavelength is 563 nm); these excitation spectra are presented in Figs. 4.36 and 4.37. It is possible to observe in intense and wide bands in fluorescence emission spectrum at 531 nm (excitation wavelength is 365 nm) and at 521 nm (excitation wavelength is 399 nm); these emission spectra are presented in Figs. 4.38 and 4.39.

Vitamin B_6 (pyridoxine) has two maxima at 367 and 400 nm in fluoresence excitation spectrum (emission wavelength is 471 nm) (Fig. 4.40), which transform into one wide spectral band at 380–400 nm if emission wavelength is 500 nm. Fluorescence emission spectrum demonstrates maximum at 470 nm (excitation wavelength is 367 nm) and at 500 nm (excitation wavelength is 400 nm); these emission spectra are presented in Figs. 4.41 and 4.42.

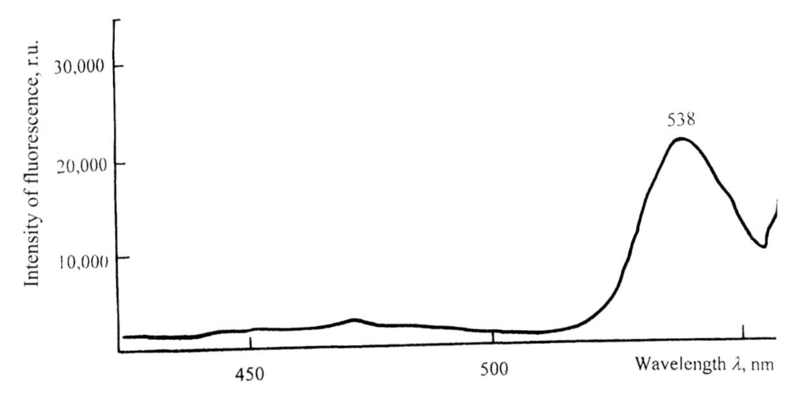

Fig. 4.37 Fluorescence excitation spectrum of vitamin B_2 (riboflavin) (emission wavelength is 563 nm)

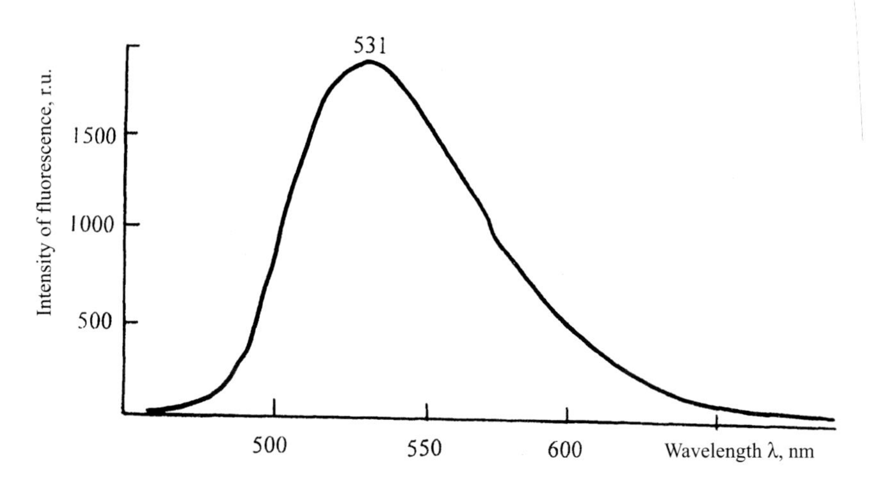

Fig. 4.38 Fluorescence emission spectrum of vitamin B_2 (riboflavin) (excitation wavelength is 365 nm)

Vitamin B_{12} (cyanocobalamin) demonstrates fluorescence excitation maximum (Fig. 4.43) at 330 nm with half-width of about 25 nm (emission wavelength is 398 nm) and fluorescence emission maximum (Fig. 4.44) at 400 nm (excitation wavelength is 345 nm).

The literature data concerning fluorescence properties of milk vitamins are as follows:

Vitamin E has maxima of fluorescence excitation at 295 nm and fluorescence emission at 340 nm [Udenfriend, 1985].

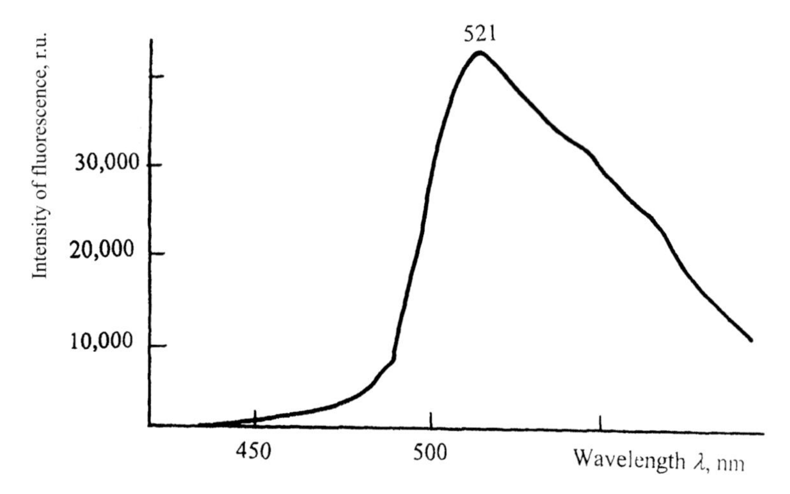

Fig. 4.39 Fluorescence emission spectrum of vitamin B_2 (riboflavin) (excitation wavelength is 399 nm)

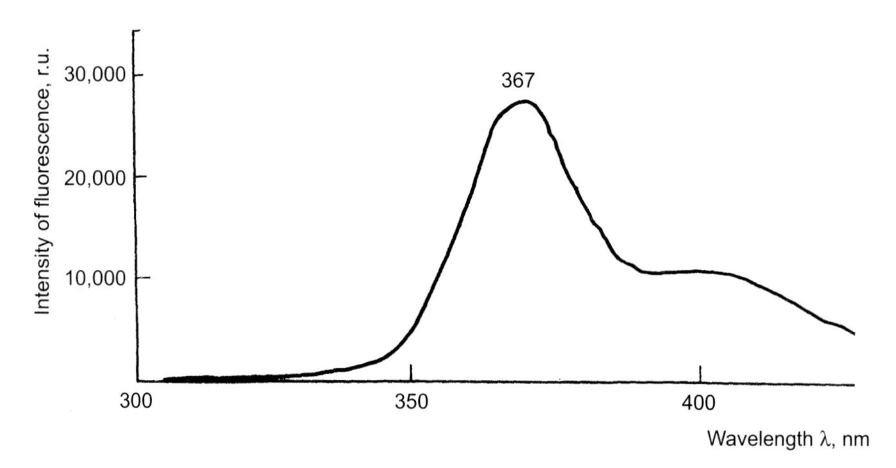

Fig. 4.40 Fluorescence excitation spectrum of vitamin B_6 (pyridoxine) (emission wavelength is 471 nm)

Vitamin B_1 (thiamine) in aqueous solutions does not demonstrate any fluorescence at room temperature [Krasnikov et al., 1987], but its oxidation product *thiochrome* shows maxima of fluorescence excitation at 365 nm and fluorescence emission at 450 nm [Vogi et al., 1955, cited by Udenfriend, 1985].

Vitamin PP has amide of nicotine acid (nicotinamide) which demonstrates fluorescence excitation maximum at 385 nm and fluorescence emission maximum at 475 nm [Krasnikov et al., 1987].

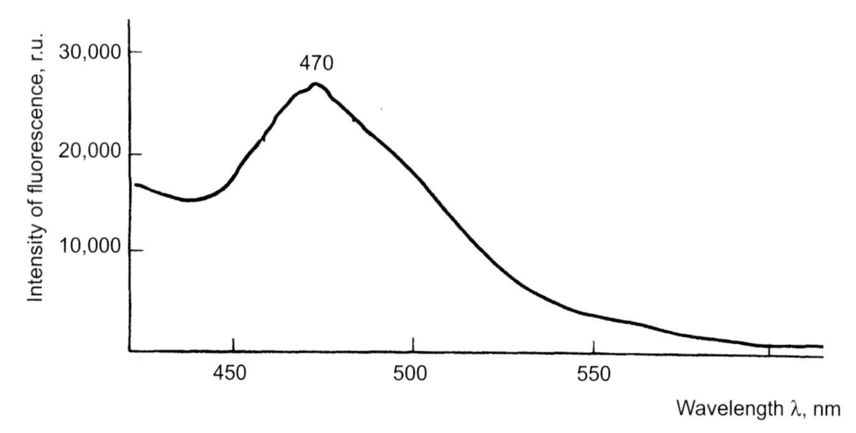

Fig. 4.41 Fluorescence emission spectrum of vitamin B_6 (pyridoxine) (excitation wavelength is 367 nm)

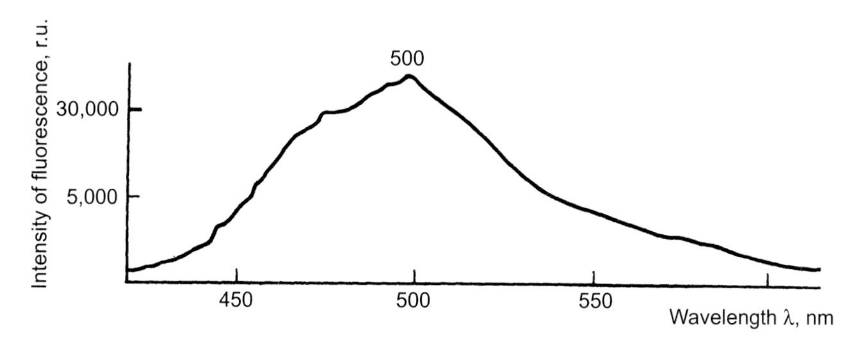

Fig. 4.42 Fluorescence emission spectrum of vitamin B_6 (pyridoxine) (excitation wavelength is 400 nm)

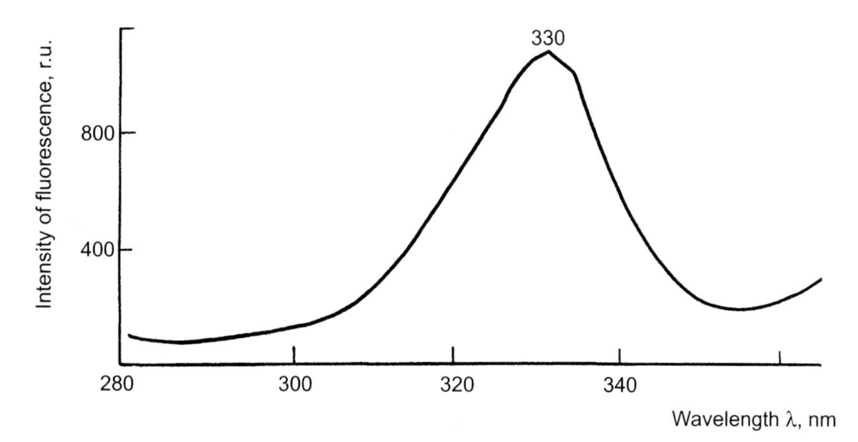

Fig. 4.43 Fluorescence excitation spectrum of vitamin B_{12} (cyanocobalamin) (emission wavelength is 398 nm)

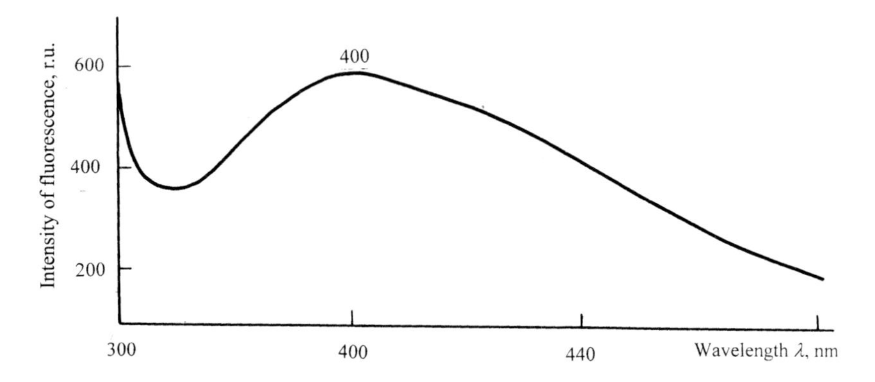

Fig. 4.44 Fluorescence emission spectrum of vitamin B_{12} (cyanocobalamin) (excitation wavelength is 345 nm)

Vitamin B_3 (pantothenic acid) is unstable in free state; there are data about fluorescence properties of the salt of this acid only - calcium pantothenate; it has fluorescence excitation maximum at 385 nm (with small maxima at 229, 260, and 385 nm) and fluorescence emission maximum at 460 nm [Krasnikov and Timoshkin, 1983; Timoshkin and Titkova, 1986].

Folic acid has fluorescence emission spectrum at 450 nm (excitation wavelength is 365 nm) and at 370 nm (excitation wavelength is 290 nm) [Daddan et al., 1957, cited by Udenfriend, 1985].

4.12 IDENTIFICATION OF MILK FLUOROPHORES

4.12.1 Selection of Excitation Wavelength

The first approach of the determination of milk components is related to the choice of the corresponding wavelength of excitation. For example, the excitation at 270–310 nm provokes the fluorescence emission of proteins and aromatic amino acids; excitation near 315–340 nm causes the fluorescence of vitamins (B_1, B_2, B_6, B_{12}, A); excitation near 340–370 nm is responsible for the emission of fat acids and vitamins (B_1, B_2, B_3, B_6, C, D, folic acid). The dependence of fluorescence emission of milk components on the excitation wavelength is presented in Table 4.10.

It is possible to vary the wavelength of excitation and to pick out a certain component (Fig. 4.45). For example, if the fluorescence of proteins at 340 nm is registered, one can be sure to avoid the influence of fat and riboflavin. There is a superposition of fluorescence emission bands of fat and riboflavin in the visible (400–700 nm) part of the spectrum; the application of interference filters makes it possible to select fluorescence of each milk component (Fig. 4.46).

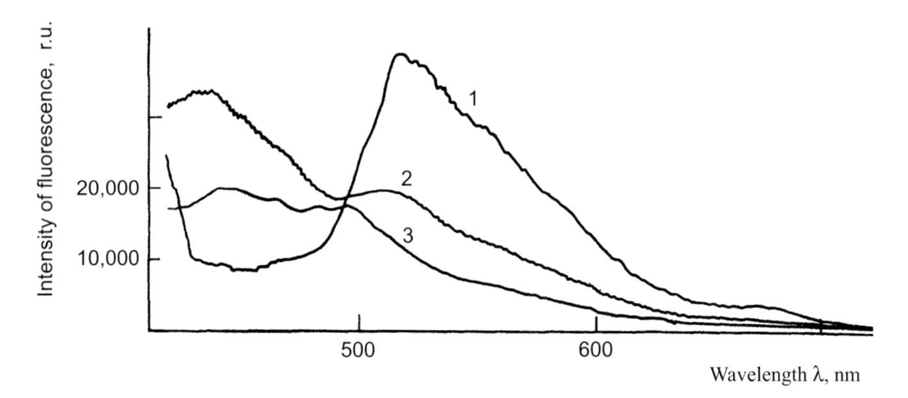

Fig. 4.45 Variation of the wavelength of excitation to pick out a certain component of milk: 1 – riboflavin; 2 – fatty acids; 3 – milk. Excitation wavelengths are: 400 nm – riboflavin; 360 nm – fatty acids; 340 nm – milk

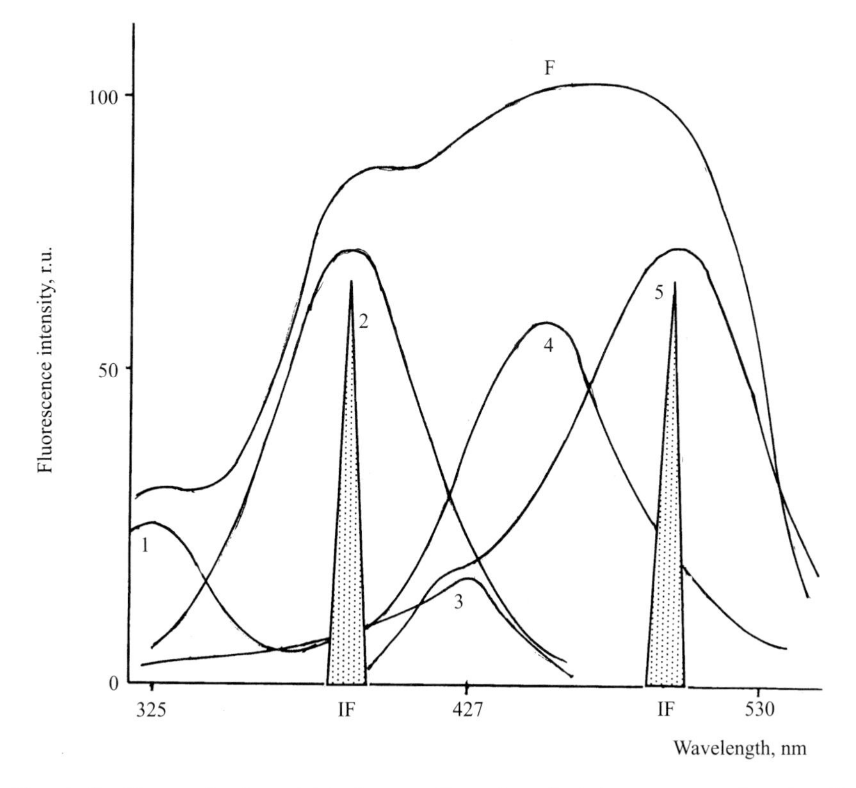

Fig. 4.46 The application of interference filters to select fluorescence of each milk component: F – spectral band of milk, 1–5 – spectral bands of milk components; IF – transmission bands of interference filters

4.12.2. Decomposition of Fluorescence Spectra

The second approach provides the determination of the fluorescence curve for a number of components; each of them is characterized by the relative intensity, wavelength, half-width and shape. The cooling of milk (up to 77 K) induces the appearance of its fine structure and facilitates the quantitative and qualitative analysis of milk components.

We have elaborated the spectra of fluorescence of milk with a universal multiparametrical and multifunctional package of programmes (UPOS) [Kucherov, 1984]. Therefore, the fluorescence emission spectrum of cooled milk (excitation wavelength is 313 nm) can be decomposed as five spectral bands (Table 4.11).

Table 4.11 Decomposition of fluorescence emission spectrum of cooled milk (excitation wavelength is 313 nm)

Band Number	*Intensity of Band (Relative Units)*	*Wavelength (nm)*	*Half-width (nm)*	*Form-Factor*		*Surface of the Band (Relative Units)*
				F_1	F_2	
1	89.24	333.0	31.58	0.41	0.60	548.03
2	257.16	381.0	51.83	0.90	0.72	2,299.71
3	66.02	409.0	27.63	0.60	0.93	323.59
4	213.79	430.0	52.35	0.69	0.20	2,368.57
5	260.55	463.0	59.47	0.06	1.00	3,531.39

It is possible to identify the following milk components such as tryptophan (N1, 333 nm), vitamin B_6 (N2, 381 nm), fat acids (N3, 409 nm), vitamin B_6 (N4, 430 nm), and vitamin C (N5, 463 nm). The results of decomposition are given in Fig. 4.47.

The fluorescence emission spectrum of cooled milk (excitation wavelength is 365 nm) can be decomposed in to two components (Table 4.12); the first components (447.52 nm) can be identified as fluorescence emission of vitamins D, B_1, and A; the second one (496.34 nm) can be associated with fluorescence emission of riboflavin.

In such a way, the procedure of decomposition of fluorescence spectra of milk is one of the perspective methods of identification of milk components. It is possible to confront parameters of spectral bands with the factors which affect the biological value of milk.

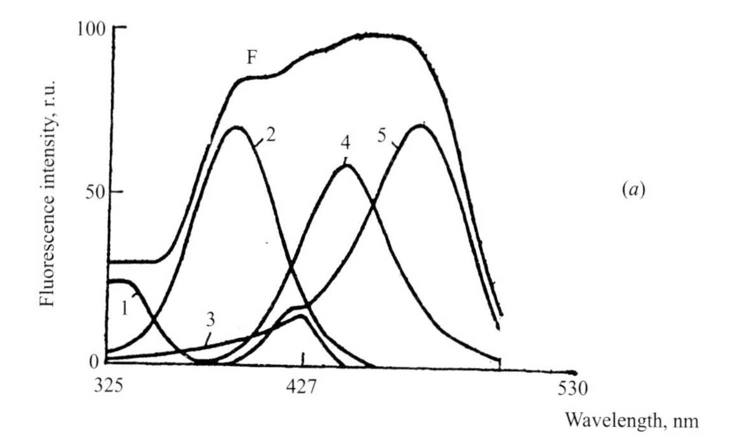

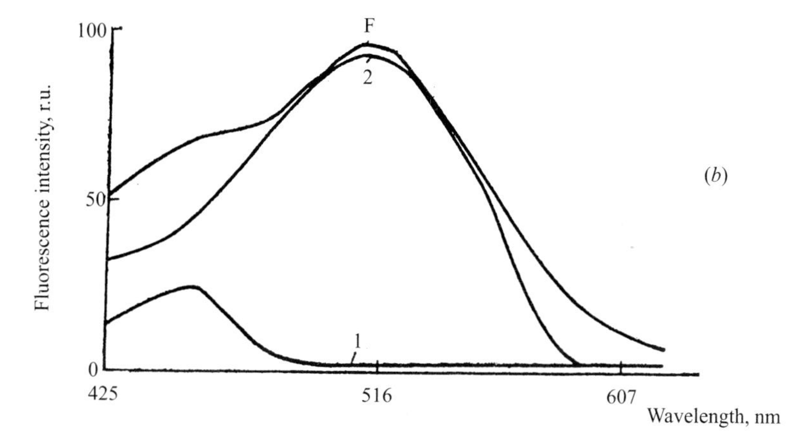

Fig. 4.47 The results of decomposition of fluorescence spectra of milk at the temperature of liquid nitrogen: (*a*) excitation wavelength is 313 nm; F – spectral band of milk, 1–5 – spectral bands of milk components; (*b*) excitation wavelength is 365 nm; F – spectral band of milk, 1–2 – spectral bands of milk components

Table 4.12 Decomposition of fluorescence emission spectrum of cooled milk (excitation wavelength is 365 nm)

Band Number	*Intensity of Band (Relative Units)*	*Wavelength (nm)*	*Half-width (nm)*	*Form - Factor*		*Surface of the Band (Relative Units)*
				F_1	F_2	
1	57.68	447.52	32.89	0.91	1.0	188.12
2	226.39	496.34	90.41	0.06	1.0	2,685.69

4.13 FLUOROMETERS FOR MILK ANALYSIS

4.13.1 Fluorometer for Determination of Protein and Fat Content in Milk

A design of fluorometer for determination of protein and fat content in milk is presented in Fig. 4.48. Mercury lamp *1* of high pressure DRK-120 was used as the source of fluorescence excitation. The radiation of this lamp was collimated by concave mirror *2*, lens *2* and *4*, and passed through interference filter *3* (280 or 325 nm), cuvette 6 with the sample (the windows of this cuvette were performed with quartz), and lens *7*. The fluorescence emission was registered at the angle 30° to the excitation radiation in order to avoid the effect of reflected and scattered radiation. The readout system consisted of the cut-off filters *8* (CZC-22 or CZC-23), and photomultiplier *9* FEU-79, amplifier *10* and recorder *11*.

The dependence of the intensity of fluorescence on the level of dilution of milk is presented in Fig. 4.49 for protein and in Fig. 4.50 for fat. This dependence demonstrates linear character in the range of concentrations 4-5 g/L.

4.13.2 Laser-induced Fluorometry of Milk

The N_2-laser fluorometer (337 nm) was used for analysis of fluorescence emission spectra of dairy products: milk, baked milk, whey and sour

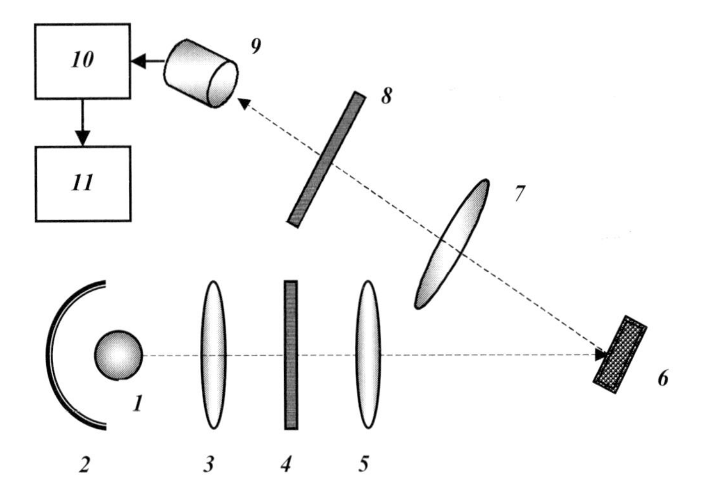

Fig. 4.48 A design of fluorometer for determination of protein and fat content in milk. 1 – mercury lamp of high pressure DRK-120 ; 2 – concave mirror; 3, 5 – lens; 4 – interference filter (280 or 325 nm); 6 – cuvette with the sample; 7 – lens; 8 – cut-off filters (CZC-22 or CZC-23); 9 – photomultiplier FEU-79; 10 – amplifier; 11 – recorder

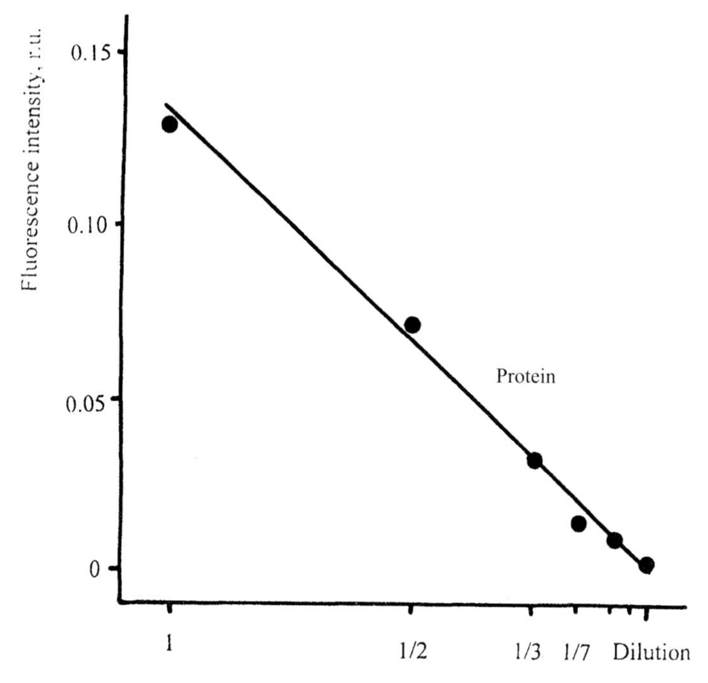

Fig. 4.49 Dependence of the intensity of fluorescence of protein on the level of dilution of milk in the range of concentrations 4–5 g/L

cream. The analysis of laser-induced spectra of milk made it possible to estimate concentrations of fat in milk samples [Posudin and Kostenko, 1994]. High level of correlation between the results which were obtained during chemical and laser-fluorometric analysis testifies in favour of laser fluorometry as method of quantitative evaluation of dairy products.

The schematic representation of laser fluorometer used [Posudin and Kostenko, 1994] for the determination of fat is given in Fig. 4.51. The radiation of nitrogen laser *1* was split by a semitransparent glass plate *2* into two beams - the first one was directed to the cuvette *3* with milk, and the second one - to fluorescence standard (solution of fluorescein) *4*. The fluorescence emission was detected in both cases with two photomultipliers *6* and *7*. The first photomultiplier was equipped with the interference filter *5* (400 nm) for selection of fluorescence emission of fatty acids. The electrical signals of both photomultipliers were directed to the differential amplifier *8* and to readout system *9*. Such a differential system can estimate the ratio of fluorescence intensities of the sample and fluorescent standard; the possible fluctuations of the laser radiation or the power supply do not affect the results of fat determination.

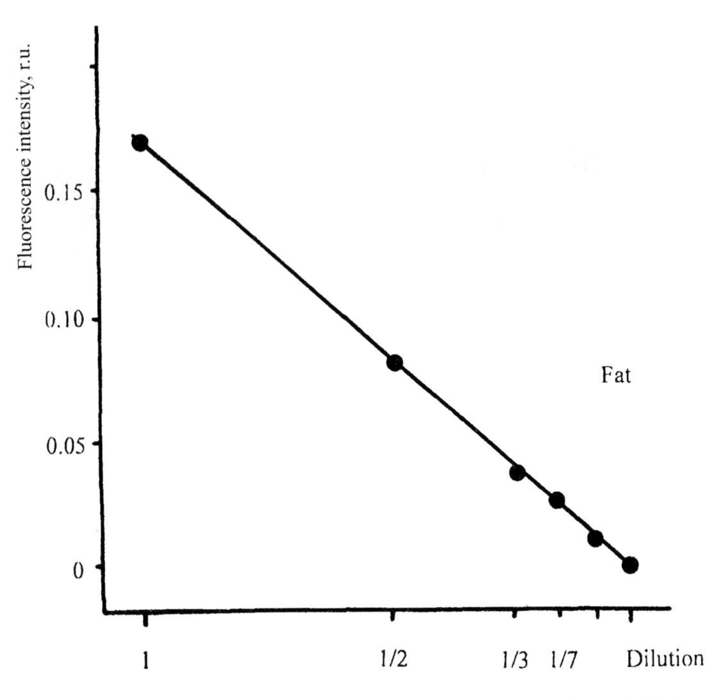

Fig. 4.50 Dependence of the intensity of fluorescence of fat on the level of dilution of milk in the range of concentrations 4–5 g/L

It is necessary to mention that the dependence of fluorescence intensity I_{Fl} on the concentration C of the sample is determined as

$$I_{Fl} = I_0 \, (1 - 10^{-\varepsilon Cl})\phi \qquad \text{(Eqn. 4.10)}$$

where I_0 is the intensity of the incident radiation, ε – molar absorption, l – optical pathlength, ϕ – quantum yield of fluorescence.

If $\varepsilon C\lambda << 0.05$, this equation can be transformed as follows:

$$I_{Fl} = I_0(2.3\varepsilon Cl)\,\phi \qquad \text{(Eqn. 4.11)}$$

Dependence of the intensity of fluorescence of milk on the level of dilution is presented in Fig. 4.52, and on the temperature in Fig. 4.53.

Thus, the fluorescence intensity depends linearly on the concentration of the sample for diluted solutions only; this method can be used in laboratory measurements.

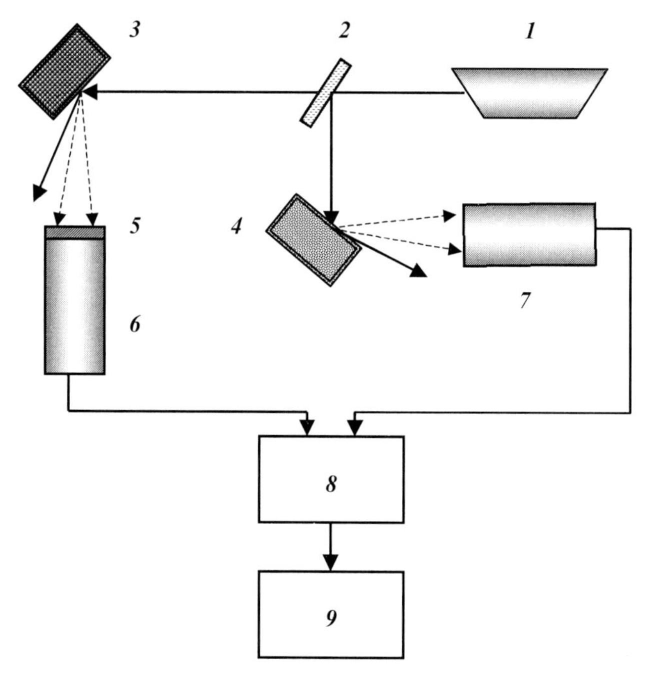

Fig. 4.51 The schematic representation of laser fluorometer used for the determination of fat [Posudin and Kostenko, 1994]. 1 – nitrogen laser; 2 – semitransparent glass plate; 3 – cuvette with milk; 4 – fluorescence standard (solution of fluorescein); 5 – interference filter (400 nm); 6, 7 – photomultipliers; 8 – differential amplifier; 9 – readout system

4.14 LASER LIGHT SCATTERING BY MILK PARTICLES

4.14.1 Theory of Light Scattering

Theoretically the process of light scattering is characterized by the so-called *coefficient of scattering* Q [Hulst, 1982; Walstra, 1964]. This coefficient depends on the wavelength of light λ, refractive indices of particles n_1 and medium n_2, and the size d of particles. The effect of these factors is determined with such parameters as relative refraction index $m = n_1/n_2$, and phase shift $\rho = 2x(m - 1)$, where $x = \pi d/\lambda$. There are different types of scattering depending on the values of all these parameters. The complete theory of light scattering is given in [Hulst, 1982]; several examples of the relation of coefficient of scattering with above-mentioned parameters are listed below.

1. *Rayleigh Scattering.* This type of scattering is demonstrated by the spherical particles which have relative refractive index equal to about 1 and dimensions less than the wavelength:

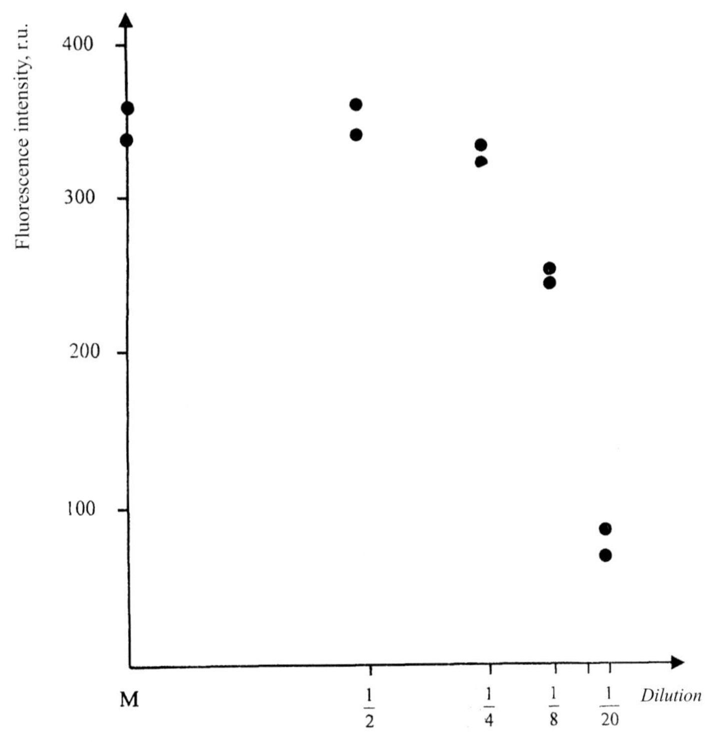

Fig. 4.52 Dependence of the intensity of laser-induced fluorescence of milk on the level of dilution of milk

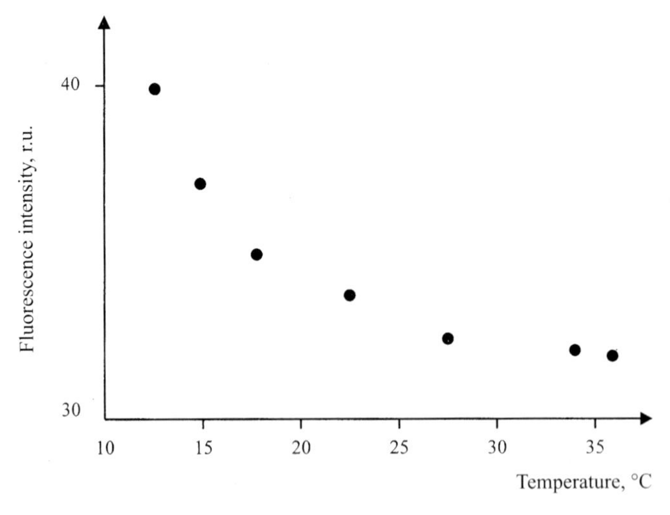

Fig. 4.53 Dependence of the intensity of laser-induced fluorescence of milk on the temperature of the sample

$$m \rightarrow 1 \quad \text{(Eqn. 4.12)}$$

$$Q_R = \frac{32}{27} x^4 (m - 1)^2 \quad \text{(Eqn. 4.13)}$$

The following expression was proposed for very small ($mx << 1$) particles [Walstra, 1965]:

$$Q_R = \frac{8}{3} x^4 \left(\frac{m^2 - 1}{m^2 + 2} \right)^2 \quad \text{(Eqn. 4.14)}$$

2. *Rayleigh-Gans Scattering*. There are two possible situations for the particles which have relative refractive index equal to about 1 and very small phase shift $2x(m - 1) << 1$:

a. Rayleigh Scattering $x << 1$; $Q_R = \frac{32}{27} x^4 (m - 1)^2$ (Eqn. 4.15)

b. Rayleigh–Gans Scattering $x >> 1$; $Q_{RG} = 2(m - 1)^2$ (Eqn. 4.16)

3. *Anomalous Diffraction*. This type of scattering is demonstrated by the spherical particles which have relative refractive index equal to about 1 and dimensions which exceed the wavelength:

$$m \rightarrow 1;\ x >> 1;\ \rho >> \lambda \quad \text{(Eqn. 4.17)}$$

$$Q_N = 2 - \frac{16m^2 \sin\rho}{(m+1)^2 \rho} + 4\frac{1 - m\cos\rho}{\rho^2} + 7.54\frac{z - m}{z + m} x^{-0.772} \quad \text{(Eqn. 4.18)}$$

where $z = [(m^2 - 1)(6x/\pi)^{2/3} + 1]^{1/2}$.

If the dimensions of particles are small or moderate, the coefficient of scattering can be written as:

$$Q_S = (1.26m - 0.44)\rho - 2.558(m - 1)^{1.273} - 0.843 \quad \text{(Eqn. 4.19)}$$

4.14.2 Application of Theory of Light Scattering to Milk Particles

Milk is a complex mixture of fat globules (0.1–10 μm), micelles of casein (0.1–0.2 μm), and particles of serum proteins (0.01–0.02 μm), and the turbidity of milk depends on the dimensions and concentration of these components. The measurement of the turbidity or optical density of milk can give information about the content of certain components. The method of light scattering seems to be one of the perspective approaches for the investigation of the size distribution of milk particles.

Table 4.13 Dependence of coefficient of scattering on the spectral region and parameters x, ρ, and d

Milk Particles	*Ultraviolet Region (λ = 0.2 μm)*			*Visible Region (λ = 0.6 μm)*			*Infrared Region (λ = 1.2 μm)*		
Fat globules $d >> \lambda$ d_{min} = 1.2 μm	15.71	3.14	Q_N	5.0	1.0	Q_N	2.62	0.52 $1.6 < \rho < 2.5$	Q_S
d_{max} = 10 μm	157.1	31.4	Q_N	49.9	10.0	Q_N	26.12	5.2 $\rho >> 2.5$	Q_N
Casein micelles $d < \lambda$ $d \sim$ 0.15 μm	2.36	0.47	Q_{RG}	0.75	0.15	Q_R	0.39	0.08	Q_R
Casein micelles $d < \lambda$ $d \sim$ 0.015 μm	0.24	0.05	Q_R	0.075	0.015	Q_R	0.04	0.01	Q_R

The dependence of coefficient of scattering on the spectral region and parameters such as x, ρ, and d is given in Table 4.13.

Dependence of the coefficient of scattering on the wavelength of light is presented in Figs. 4.54 and 4.55 for milk globules, and in Figs. 4.56 and 4.57 for casein micelles and serum protein particles. Thus, the analysis of the dependence of light scattering on the wavelength and angle of

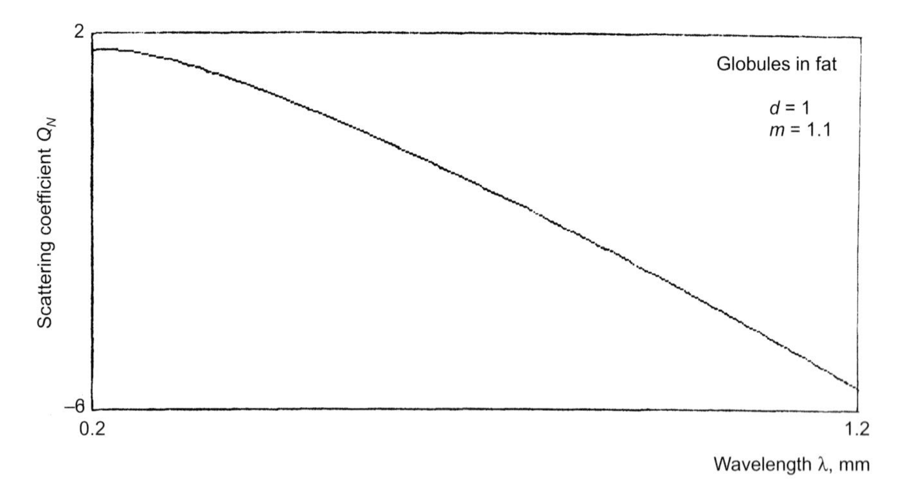

Fig. 4.54 Dependence of the coefficient of scattering on the wavelength of light for milk globules (diameter 1 μm)

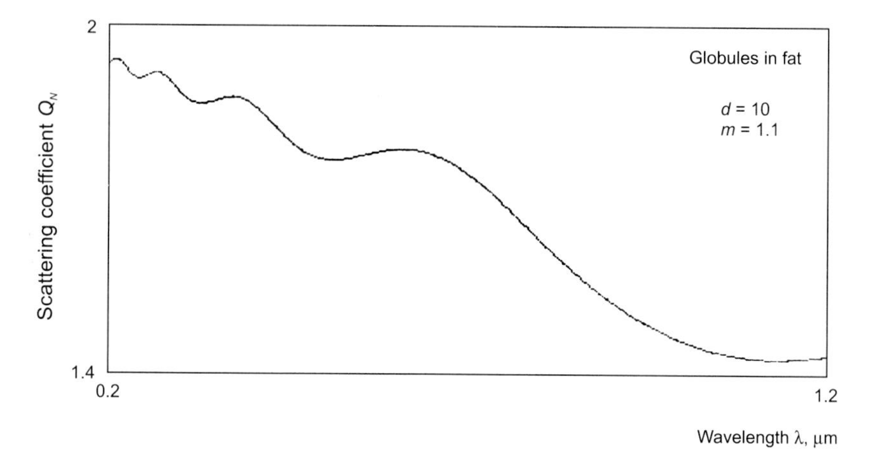

Fig. 4.55 Dependence of the coefficient of scattering on the wavelength of light for milk globules (diameter 10 μm)

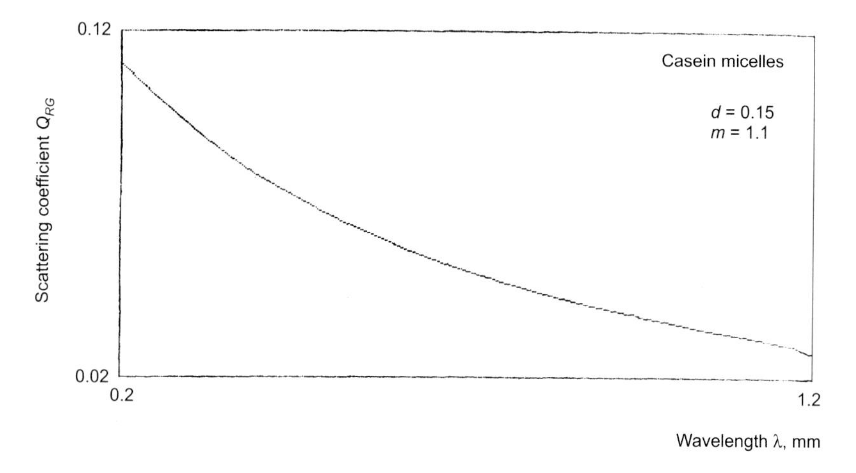

Fig. 4.56 Dependence of the coefficient of scattering on the wavelength of light for casein micelles

observation can make it possible to estimate quantitatively the dimension of milk particles.

It is clear that the transfer from visible to infrared part of the spectrum induces the decreasing of coefficient Q_N 1.35 times for fat globules of diameter 10 μm and 4.41 times for fat globules of diameter 1 μm. That is why it is expedient to pass to the infrared region in order to determine the fat content in milk and to avoid the effect of small particles.

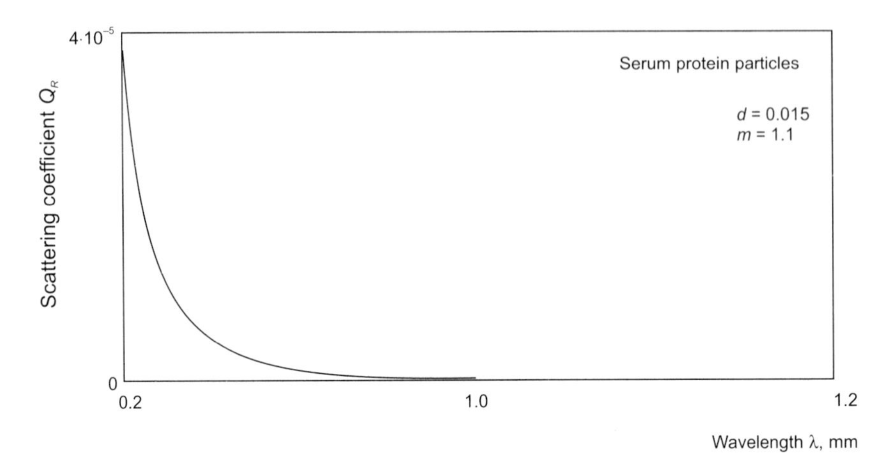

Fig. 4.57 Dependence of the coefficient of scattering on the wavelength of light for serum protein particles

4.15 INSTRUMENTATION

The experimental investigation of light scattering by milk particles was done with the photon correlation spectrometer *System 4700c*. The schematic representation of this spectrometer is presented in Fig. 4.58. The focused radiation of laser *1* is directed to the sample *2* of milk that is placed on the goniometer table *3*. The scattered radiation is directed to the readout system which consists of photomultiplier *4*, correlator *5*, computer *6*, and printer *7*. A sample holder is supplied with a system of temperature control *8*, peristaltic pump *9*, and electric motor *10*. A signal from the output of photomultiplier gives the information about the dependence of light scattering on the angle of observation. Milk, aqueous solution of lactose, dry milk, and milk mixtures were used as the samples.

4.16 RESULTS OF LIGHT SCATTERING BY MILK PARTICLES

The typical histograms of size distribution of these samples are given in Figs. 4.59–4.63. It is clear that the maximum of size distribution of milk particles is shifted with increasing concentration of particles from 100–500 nm ($C \leq 10^{-3}\%$) to 500–2,000 nm ($C = 7.4\%$). The character of size distribution of milk particles depends strongly on the concentration (Fig. 4.59–4.62) and the heating of the sample (Figs. 4.62–4.63).

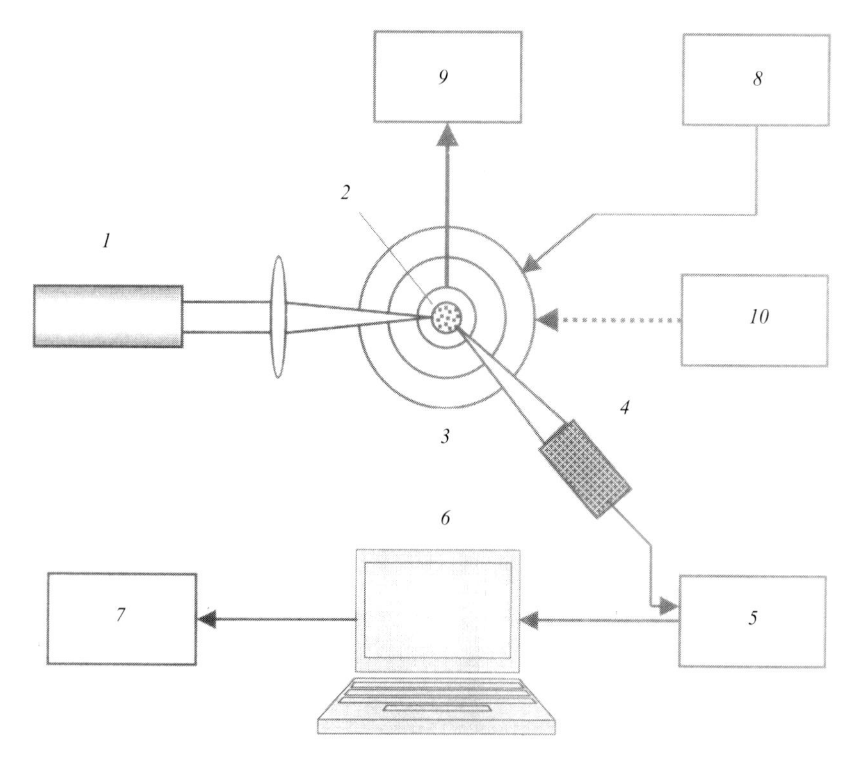

Fig. 4.58 Schematic representation of photon correlation spectrometer *System 4700c*. 1 – laser; 2 – sample; 3 – goniometer table; 4 – photomultiplier; 5 – correlator; 6 – computer; 7 – printer; 8 – system of temperature control; 9 – peristaltic pump; 10 – electric motor

Fig. 4.64 demonstrates the dependence of the intensity of scattered light on the angle of observation.

In such a way, the method of light scattering can be used for analysis of the effect of milk content and technological processes which are related to heating (pasteurization, sterilization, homogenization) on the size distribution of milk particles.

4.17 SUMMARY

Method of absorption/transmission spectroscopy of milk in the visible part of spectrum demonstrates a serious limitation - the milk sample should be adequately diluted to obtain the linear dependence of optical density on the concentration of milk sample which must be evaluated.

Method of infrared microspectrophotometry of milk is based on the measurement of the absorption of the main components of milk (fat,

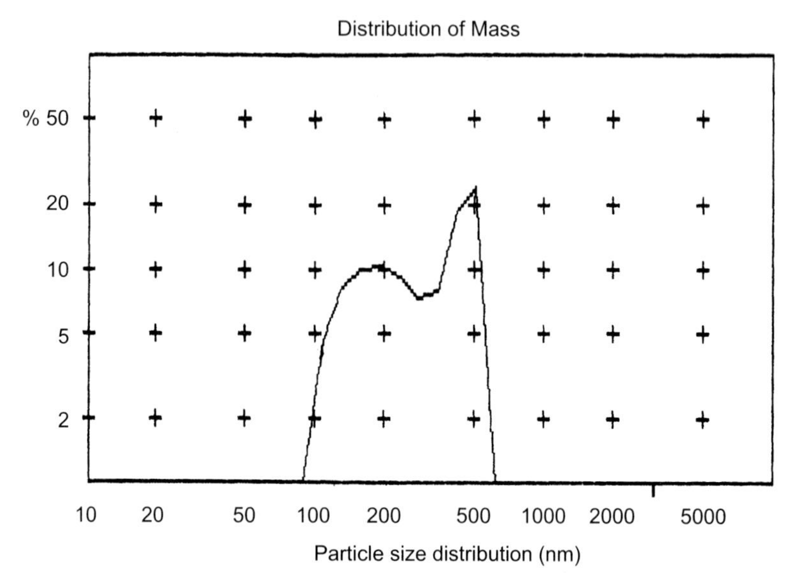

Fig. 4.59 Typical histogram of size distribution of the diluted (≤ 10^{-3}%) milk. Temperature 25°C

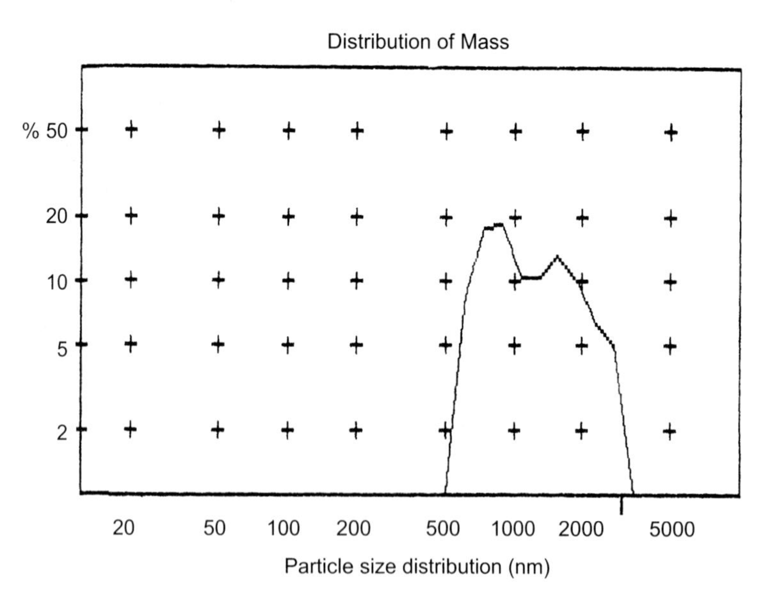

Fig. 4.60 Histogram of size distribution of milk. Concentration of fat 4.2%; temperature 20°C

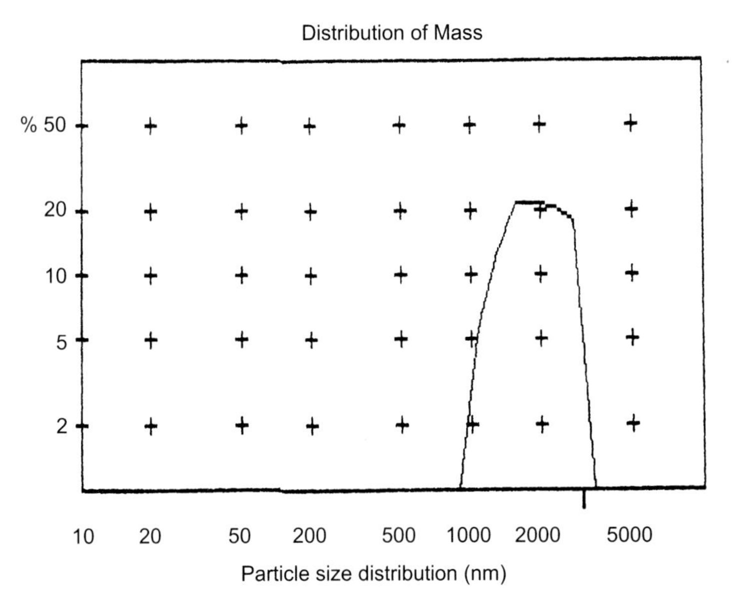

Fig. 4.61 Histogram of size distribution of milk. Concentration of fat 5.2%; temperature 20°C

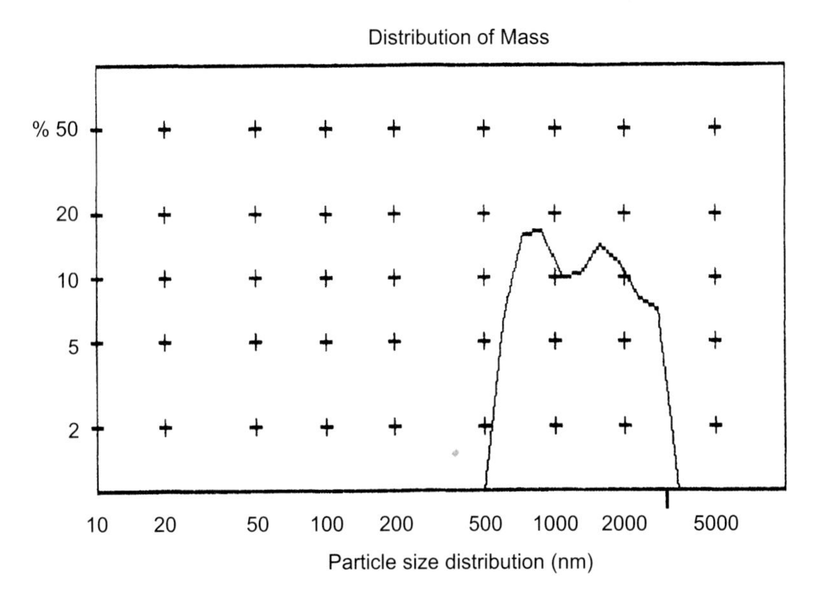

Fig. 4.62 Histogram of size distribution of milk. Concentration of fat 7.4%; temperature 20°C

protein, lactose) and quantitative evaluation of these components. The spectral position of infrared absorption bands depends on the temperature, level of milk homogenization, and pH of the milk.

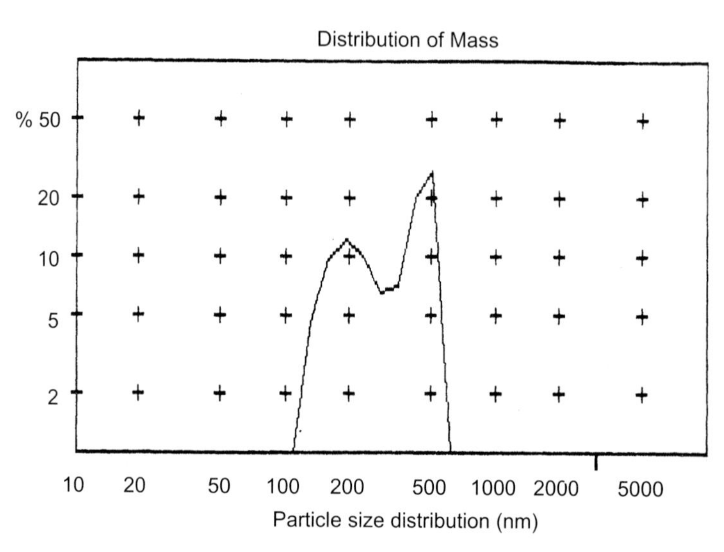

Fig. 4.63 Histogram of size distribution of milk. Concentration of fat 7.4%; temperature 50°C

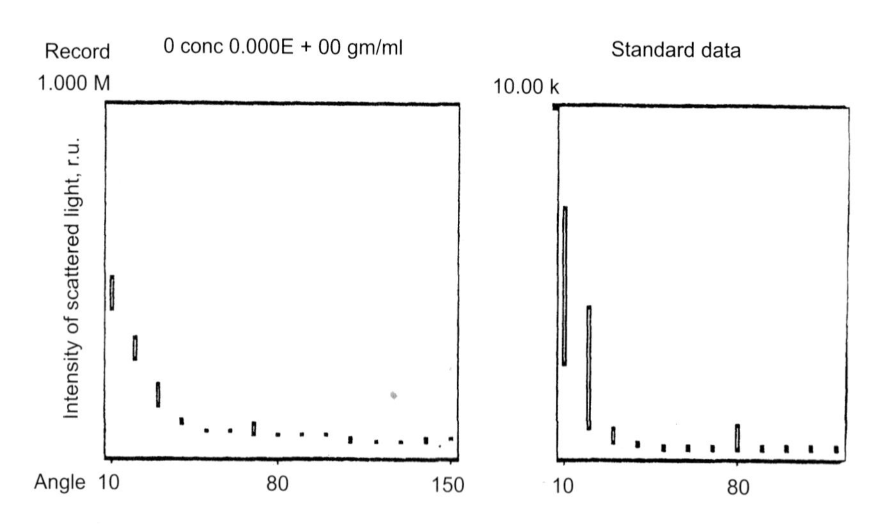

Fig. 4.64 Dependence of the intensity of scattered light on the angle of observation

A number of milk components (proteins, fat, vitamins) demonstrate fluorescence. However, the fluorescence intensity of milk depends linearly on the concentration of the sample for diluted solutions only; the fluorescence method can be used in laboratory measurements.

The method of near-infrared spectroscopy provides the possibility to analyze milk samples which contain a high proportion of water and demonstrate high level of opacity. The NIR method is rapid, non-destructive and offers a high level of accuracy in comparison with traditional chemical methods.

The method of light scattering can be used for analysis of the size distribution of milk particles and the effect of technological processes which are related to heating (pasteurization, sterilization, homogenization) on this distribution.

4.18 REFERENCES

Biggs, D.A. and Sjaunja, L.O. 1987. Analysis of Fat, Protein, Lactose and Total Solids by Infra-Red Absorption. *Bull. Int. Dairy Food.* 208: 21-30.

Brusilovsky, L.P. and Waynberg, A.Ya. 1990. *Devices of Technical Control in Dairy Industry.* Moscow. p 287.

Fuhrman, I. 1987. Zur anwendung der nah-infrarotreflexions-spektroskopie in der milchwirtschaft. *Lebensmit teilidustrie.* 4: 171-174.

Goulden, J.D.S. 1964. Analysis of milk by infrared absorption. *J. Dairy Res.* 31: 273-284.

Hulst, H.C. 1982. *Light Scattering by Small Particles.* Dover Publ. Inc. p 470.

Krasnikov, V.V. and Timoshkin, E.I. 1983. *Luminescence of Foods.* Moscow, Light and Food Industry. p 264.

Krasnikov, V.V. and Timoshkin, E.I. and Titkova, A.V. 1987. *Spectral Luminescence Analysis of Foods.* Moscow, Agropromizdat. p 288.

Kucherov, A.P. 1984. Calculation of the values which characterise the overlapped spectral lines. *J. Applied Spectroscopy.* 41(1): 79-82.

Martynenko, I.I., Posudin Yu.I. and Shevchenko A.I. 1996. Infrared photometer. *Food and Processing Industry.* 2: 30.

Posudin, Yu.I. 1988. *Lasers in Agriculture.* Science Publishers, Inc. Enfield, NH, USA. p 188.

Posudin, Yu.I. 1993. Luminescence method of milk control. *Isvestia VUZov. Food Technology.* 5-6: 79-81.

Posudin, Yu.I. 1993. Turbidimetric methods of milk control. *Isvestia VUZov. Food Technology* 3-4: 85-87.

Posudin, Yu.I. 1995. Laser fluorometer for determination of fat in milk. *Isvestia of VUZov. Food Technology* 3-4: 72-73.

Posudin, Yu.I. 1996. *Methods of Optical and Laser Spectroscopy in Agriculture.* Seminar Series of visiting Fulbright Scholar. University of Georgia, Athens, U.S.A. January-April 1996.

Posudin, Yu.I. 2002. *Method of Determination of Carbohydrates in Biological Objects.* Patent of Ukraine, 29114A, N 98010144, 2002.

Posudin, Yu.I. 2005. *Methods of Nondestructive Quality Evaluation of Agricultural and Food Products.* Aristey, Kyiv. p 408.

Posudin, Yu.I. and Kostenko, V.I. 1989. Laser determines fat content. *Agroprom of Ukraine.* 4: 64.

Posudin, Yu.I. and Kostenko, V.I. 1991. Method of laser spectrofluorometry of milk and dairy products. *Agr. Biology.* 2: 191-195.

Posudin, Yu.I. and Kostenko, V.I. 1991. *Method of determination of fat content in milk.* Invention N 1698715 USSR.

Posudin, Yu.I. and Kostenko, V.I. 1992. Determination of milk content on the basis of Infrared spectrophotometry. *Isvestiya VUZov. Food Technology.* 3-4: 64-66.

Posudin, Yu.I. and Kostenko, V.I. 1994. *Spectroscopy of Milk.* Urozhaj, Kyiv. p 83.

Posudin, Yu.I., Martynenko, I.I. and Shevchenko, A.I. *Instrument for Quantitative Determination of Milk Components.* Patent of Ukraine, 14626 A. N95041858, 1997.

Posudin, Yu.I. and Timoshenko Yu. A. 1991. *Modern Methods of Determination of Milk Content.* Agr. Academy Publ., Kiev. p 69.

Timoshkin, E.I. and Titkova, A.V. 1986. Spectral characteristics of luminescence of several vitamins. *Moscow. Depon. VINITI.* N 1551. B86.

Udenfriend, C. 1985. *Fluorescence Analysis in Biology and Medicine.* Moscow, Mir. p 484.

Voronkova, E.M., Grechushnikov, B.M., Distler, G.I. and Petrov, I.P. 1965. *Optical Materials for Infrared Technique.* Nauka, Moscow, Nauka. p 322.

Walstra, P. 1964. Approximation formulae for the light scattering coefficient of dielectric spheres. *J. Appl. Puegs.* 15(12): 1545-1552.

Walstra, P. 1965. Light scattering by milk fat globules. *Neth. Milk Dairy J.* 19(2): 93–109.

CHAPTER

5

Spectroscopic Analysis of Eggs

5.1 PROPERTIES OF EGGS

5.1.1 Structure and Contents of the Egg

The main parts of the egg are presented in Fig. 5.1. In the centre of the egg is the yolk which consists of latebra, germinal disc (blastoderm), and concentric light and dark layers, surrounded by the vitelline membrane. The albumen consists of the inner thick white (chalaziferous), inner thin white, thick white and outer thin white layers. The next layers are the inner and outer shell membranes. The thickness of these layers is about 0.01–0.02 mm [Stadelman and Cotterill, 1973]. All these elements are covered with the eggshell, which includes the cuticle, spongy and mammillary layers.

The main constituents of albumen are water (86–88%) and total solids (12–14%) which consist of proteins (9.7–10.6%).

The yolk has 50% total solids, which include proteins (15.7–16.6%) and lipids (31.8–35.5%).

The eggshell consists of calcium carbonate (94%), magnesium carbonate (1%), calcium phosphate (1%) and organic matter (4%).

5.1.2 Eggshell Colour and Pigmentation

The colour of the eggshell can be explained as a means of camouflage and temperature control [Solomon, 1997]. The main pigments which are responsible for the colouration of the eggshell are shell *porphyrins* or *ooporphyrins* – cyclic compounds formed by the linkage of four pyrrole rings through methylene bridges. These pigments are presented by the *protoporphyrin*, *uroporphyrin* and *coproporphyrin*.

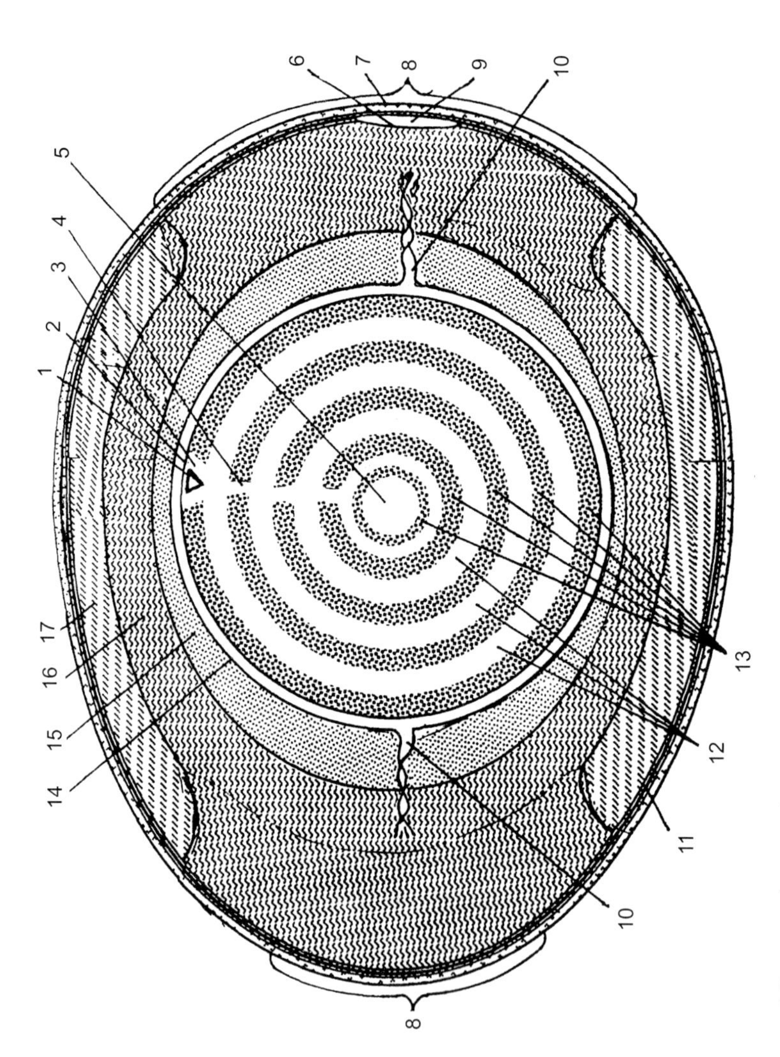

Fig. 5.1 Structure of the egg: 1 – germinal disc (blastoderm); 2 – yolk (vitelline) membrane; 3 – Pender's nucleus; 4 – latebra's neck; 5 – latebra; 6 – egg membrane; 7 – eggshell; 8 – ligamentum albuminis; 9 – air cell; 10 – chalazae; 11 – shell membrane; 12 – light yolk layer; 13 – dark yolk layer; 14 – chalaziferous; 15 – inner fluid layer of albumen; 16 – dense layer of albumen (albuminous sac); 17 – outer liquid layer of albumen

5.2 CURRENT PROBLEMS OF EGG QUALITY EVALUATION

The modern poultry industry is not satisfied with the traditional system of handling and processing of eggs, which is based on candling - visual inspection of the eggs. Currently, the operator of the conveyor does not have the opportunity to inspect 25,000 eggs per hour and estimate the freshness, weight, bacterial infection, presence of technical spoilage, and eggshell defects without elimination of subjectivity, fatigability and destruction. That is why automatization of egg quality control is rather actual.

Measures of egg quality. The main parameters of the egg which are inspected during its quality evaluation are: freshness of the egg, presence of the interior quality defects, weight, density and shape (egg index, asymmetry) of the egg, state of the eggshell, size of the air cell, albumen and yolk quality, Haugh unit, ratio of albumen weight to yolk weight, and the eggshell thickness.

Main interior quality defects in eggs. The most common defect found in eggs is related to *blood spots*; another type of internal spoilage is *meat spots*. The nature of blood and meat spots and factors are discussed by Nalbandow and Card (1944), and Helbaska and Swanson (1958). A less frequent defect is *yolk mottling*, which means a non-uniform distribution of water [Polin, 1957] or a separation of the vitelline membrane and the chalaziferous layer of the albumen [Doran and Muellar, 1961].

Eggshell defects. The most commonly detected surface defects and irregularities are [Solomon, 1997]: calcium splash, fine and heavy dusted, chalky deposit, lilac and pink eggs, body-cheched egg (equatorial bulge), white-banded and slab-sided eggs, corrugated eggs, and various mechanical defects (cracks, breakings, incisions).

5.3 ABSORPTION/TRANSMISSION SPECTROSCOPY OF EGGS IN VISIBLE PART OF SPECTRUM

Instrumental measurement of light transmission can be considered as an alternative to visual candling of the eggs. Spectrophotometry of shell colour [Brant et al., 1953], yolk colour [Philip et al., 1976], blood and meat spots [Brant, 1953] present the first attempt of the automation of egg quality control. The principle of blood and meat spots detection was based on the measurement of light transmission at 575 μm where haemoglobin has the absorption band [Blackburn, 1964]. The disadvantage of this method is the dependence of the transmitted light on the size, colour of egg, sickness and contamination of the eggshell. The detection of the technical spoilage of the eggs on the basis of transmission spectra analysis was proposed by Mayorov et al. [In: Tzarikov and Chernova, 1968]. It was shown that fresh white eggs had the maximum transmission at 573–580 nm, brown eggs at 590–596 and 636–638 nm, and eggs with blood spots had absorption maxima at 538–540 and 573–576 nm.

The electronic ovoscope for egg quality grading was developed in a United Kingdom packing station [Wells and Belyavin, 1987]. This system consisted of high-resolution cameras which were capable of inspecting 48 eggs moving on a conveyor. The picture which was projected on a screen, allowed the operator to electronically label the suspected egg and remove it from the conveyor.

It is very attractive to use the spectrophotometry of eggs for the nondestructive determination of the fecundated eggs. The fact is that fecundated and nonfecundated eggs differ with the production of carbon dioxide as the result of egg metabolism (threefold according to Duley [In: Tzarikov and Chernova, 1968]. However, carbon dioxide is responsible for the absorption at 426.8 nm; the measurement of the absorption of CO_2 can make it possible to determine the fecundated eggs and to select them. Unfortunately, this problem has not yet been resolved.

5.4 FLUORESCENCE SPECTROSCOPY OF EGGS

The fluorescence characteristics of the eggs present a convenient quality criterion. Bogolubsky (1958) was the first to propose the application of visual estimation of eggs fluorescence to their selection; he introduced eight classes of colouration of eggshell during fluorescence emission.

The excitation of fluorescence by ultraviolet radiation was used for the quality evaluation of eggs with different levels of pigmentation [Podlegayev and Sarycheva, In: Tzarikov and Chernova, 1968; Van Deren and Jaworski, 1968]. The possibility to distinguish the eggs which were infected by *Pseudomonas* was also demonstrated.

The dependence of the fluorescence of eggs on the season, hen age, egg productivity and impregnation was investigated by Bekhtina and Dyagileva (1968).

The fluorescence excitation (with maxima at 405, 510, 540, and 575 nm) and emission (635 and 700 nm) were related to the pigments of the porphyrin nature – ptotoporphyrin IX and porphyrin derivatives of florin and oxoflorin [Rybalova and Maslova, 1985].

The detection of spoiled eggs on the basis of fluorescence measurements was proposed by Imai and Saito (1985). The intensity of fluorescence at 457 and 490 nm were used as spectral criteria of egg quality.

5.5 DEFECTOSCOPY OF EGGSHELL

The lighting methods (strobe and incandescent) were used for the acquisition of images and for the detection of cracks in eggs [Goodrum and Elster, 1992]. Strobe lighting appeared to be an acceptable alternative to incandescent light for machine vision inspection of light.

The laser scanning system as a means of detecting hair cracks in eggs was developed by Bol (1981).

The method of laser defectoscopy of eggs on the basis of the Faraday effect was proposed by Bezverkhny and Pigarev (1984). The egg on the conveyor is illuminated by laser radiation. Either incident or reflected radiation is polarized. The external magnetic field which is applied to the egg can rotate the plane of polarization of the reflected radiation. The angle of rotation depends on the state of eggshell, particularly on the mechanical defects. The system can respond to the intensity of reflected radiation and provide selection of the eggs.

Nondestructive evaluation by the holographic method of the strength of an egg was proposed by Vikram et al. (1980). The analysis of interference between reflected and reference laser beams made it possible to estimate the deformations and the strength of eggshell through the calculation of the interference bands.

5.6 INSTRUMENTATION

The absorption spectrometer consisted of a halogen lamp, monochromator, sample, photomultiplier and readout system. A signal $A = (I - I_N)/(I_0 - I_N)$ was registered, where signals I and I_0 were recorded with and without egg correspondingly, and I_N was a level of noise which was estimated with closed photomultiplier.

Fluorescence properties of the eggshell were evaluated with luminescence microscope LUMAM-3M. The mercury lamp DRSh-250 *1* was used as the source of fluorescence excitation; the radiation of this lamp passed through water infrared filter *2*, glass filter *3*, semitransparent plate *4*, and objective *5* of the microscope on the sample *6*. Fluorescence radiation was registered through the plate 4 and filter *7* by photomultiplier FEU-79 *8* and readout system H–306 *9* (Fig. 5.2).

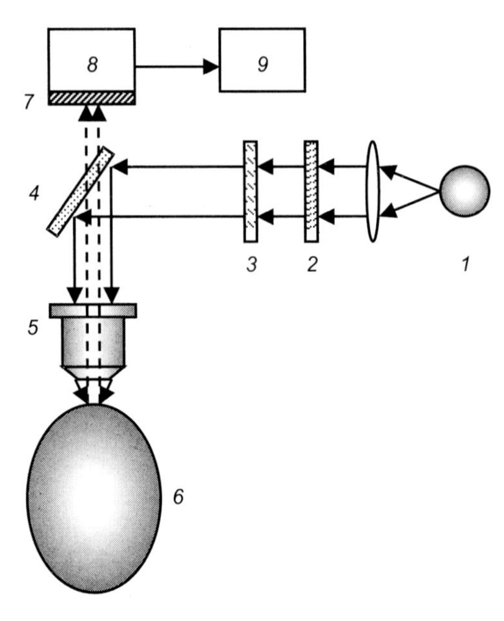

Fig. 5.2 An optical design of fluorometer for studying the fluorescence properties of the eggshell. 1 – mercury lamp DRSh-250; 2 – water infrared filter; 3 – glass filter; 4 – semitransparent plate; 5 – objective of the microscope; 6 – sample; 7 – filter; 8 – photomultiplier FEU-79; 9 – readout system H-306 [Posudin and Lepeshenkov, 1991; Posudin et al., 1992]

Fluorescence decay is an important parameter which can also be used for the quantitative estimation of eggshells. It was shown that the fluorescence decay process is described by two-exponential function [Posudin and Lepeshenkov, 1991; Posudin et al., 1992]:

$$I(t) = Ae^{-t/T_1} + Be^{-t/T_2} \quad \text{(Eqn. 5.1)}$$

where $I(t)$ is intensity of fluorescence, A and B are constants; T_1 and T_2 - decay time of both exponential curves, t - time, e - base of the natural logarithm.

Laser defectoscopy of the eggshell was based on the interaction of laser radiation that passes into objective of the microscope through the eggshell cracks during rotation of the egg. The reflected laser beam will change its intensity and direction; alterations of the reflected beam give the information about crack parameters. The typical laser defectoscope consists of laser 1, mirror 2, microscope objective 3, egg sample 4, photodetector 5, and readout system 6 (Fig. 5.3).

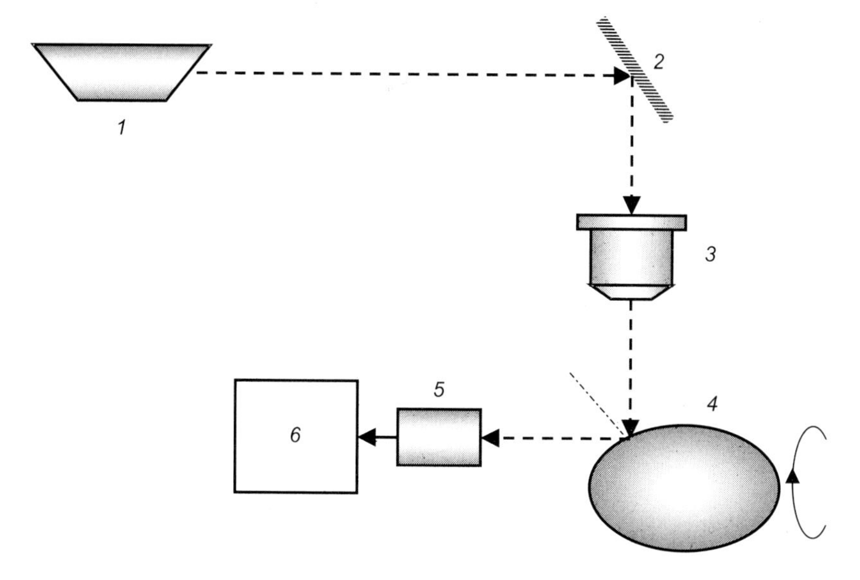

Fig. 5.3 Principles of laser defectoscopy of the eggshell. 1 – laser; 2 – mirror; 3 – microscope objective; 4 – egg sample; 5 – photodetector; 6 – readout system [Posudin and Lepeshenkov, 1991]

5.7 RESULTS OF SPECTROSCOPIC ANALYSIS OF EGGS AND EGGSHELLS

5.7.1 Reflectance Spectra

The dependence of factor such as $1 - I_{sh}/I_{st}$ on the wavelength λ was estimated as reflectance spectrum of the eggshell of white and brown colour, where I_{sh} and I_{st} are intensities of light reflected from the eggshell and standard (chalky paper) correspondingly. These reflectance spectra which demonstrate the maxima at 480, 535, 563, 590 and 642 nm, are presented in Fig. 5.4 [Posudin and Lepeshenkov, 1991, 1992]. The intensity of reflected light is different for white and brown eggshells.

5.7.2 Transmission Spectra

The typical spectra of transmission of the hen and duck eggs are presented in Fig. 5.5 [Posudin et al., 1992, 1995, 1998, 2005]. It is possible to distinguish the spectra of the fresh eggs (1), eggs with blood inclusions (2), and spoiled eggs (3). The transmission peak for fresh white hen egg is located at 573–580 nm; the spectra of cream fresh eggs have two peaks at 590–596 nm and 636–638 nm. It is clear that the spectral position of transmission maximum of eggs at 550–600 nm

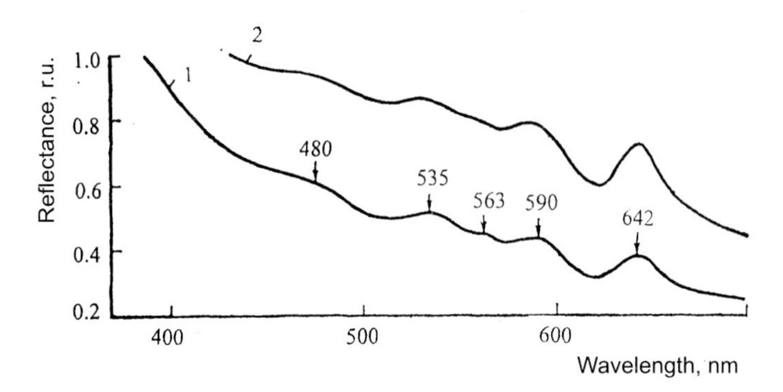

Fig. 5.4 Reflectance spectra of eggshell: 1 – light cream colour; 2 – dark cream colour [Posudin and Lepeshenkov, 1991, 1992]

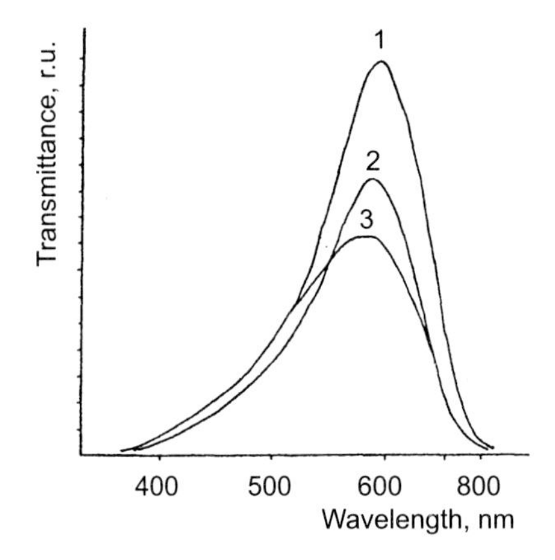

Fig. 5.5 The typical spectra of transmission of the hen and duck eggs. 1 – fresh eggs; 2 – eggs with blood inclusions; 3 – spoiled eggs [Posudin and Lepeshenkov, 1991, 1992; Posudin et al., 1992]

depends on the freshness of the eggs and presence of technical spoilage [Posudin and Lepeshenkov, 1991, 1992; Posudin et al., 1992].

5.7.3 Fluorescence Spectra

The fluorescence spectroscopy is a rather promising approach for the detection of different types of spoilage and quantitative estimation of eggshell pigmentation [Posudin et al., 1992]. The typical spectra of fluorescence emission of hen and duck eggs, fresh and spoiled, with different eggshell colour are presented in Fig. 5.6. The intensity of the main maximum at 672–674 nm depends on the freshness of the eggs, level of eggshell pigmentation, and the interior defects.

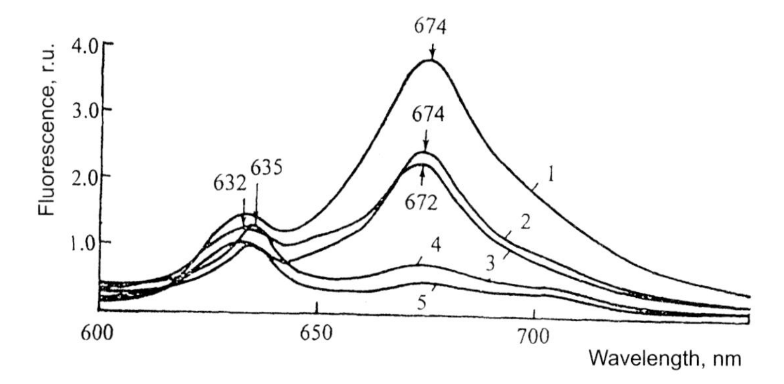

Fig. 5.6 Typical spectra of fluorescence emission of hen and duck eggs, fresh and spoiled, with different eggshell colour. 1-3 – hen eggs; 4-5 – duck eggs; 1 – dark cream colour; 2, 3 – light cream colour; 5 – eggs with blood inclusions [Posudin et al., 1992]

The typical fluorescence decay curves for the eggs with different levels of pigmentation are presented in Fig. 5.7 [Posudin et al; 1992; Posudin and Kucherov, 1992]. Here the level of pigmentation is decreasing from N1 to N7. The quantitative data of the fluorescence decay are presented in Table 5.1.

Table 5.1 Amplitude and temporal constants of exponents

Amplitude of Exponents		*Temporal Constants of Exponents*	
A	*B*	τ_1	τ_2
82.297	21.181	636.224	17.567
81.932	41.596	621.063	38.838
66.256	32.713	413.558	14.777
45.246	22.350	629.568	21.525
92.594	43.872	598.981	26.776
60.115	49.165	369.775	10.948
35.627	33.817	216.245	7.665

5.7.4 Defectoscopy of Eggshell

Laser radiation with a small divergence can be applied to the detection of eggshell cracks. The minimal spot diameter of laser radiation after its focusing by a microscope is determined by the numerical aperture, wavelength of light and focal distance of the objective [Posudin and Lepeshenkov, 1991, 1992; Posudin et al., 1992]:

$$D_{\min} = 2\omega f. \qquad \text{(Eqn. 5.2)}$$

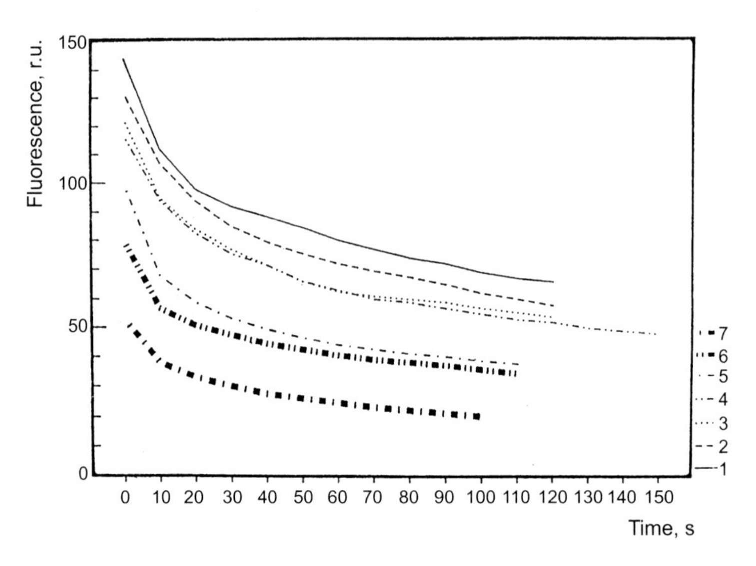

Fig. 5.7 Typical fluorescence decay curves for the eggs with different levels of pigmentation. Here the level of pigmentation is decreasing from N1 to N7 [Posudin et al., 1992; Posudin and Kucherov, 1992]

Combination of He:Ne laser ($\omega = 5{\cdot}10^{-4}$ rad, $\lambda = 632$ nm, $P = 55$ mW) with the objective 8 × 0.20 (focal distance $f = 18.2$ mm) gives

$$D_{min} = 2.5{\cdot}10^{-4}\, 18.2{\cdot}10^{-3} = 18.2\ \mu\text{m}$$

and with the objective 40 × 0.65 (focal distance $f = 4.25$ μm)

$$D_{min} = 2.5{\cdot}10^{-4}\, 4.25{\cdot}10^{-3} = 4.25\ \mu\text{m}$$

These values of D_{min} can be considered as threshold diameters of laser radiation which can be realized during defectoscopy of the eggshell.

Further progress in the egg quality practice is related to the development of automated instrumentation of control, such as image analysis system. The main procedure in these systems is the scanning of the object to assess its size and shape. The information is elaborated by a microprocessor and is used for the sorting of eggs.

5.8 SUMMARY

Spectroscopic methods can be applied to non-destructive evaluation of poultry-farming products. The intensity of reflected light is different for white and brown eggshells. The maximum spectral position of transmission of eggs depends on the freshness of the eggs and presence of technical spoilage. The intensity of fluorescence maxima depends on the freshness of the eggs, level of eggshell pigmentation, and the interior defects. Laser radiation with a small divergence can be applied to the detection of eggshell cracks.

5.9 REFERENCES

Bekhtina, V.G. and Dyagileva, G.E. 1968. Fluorescence of eggs and economically useful signs of hens. In: *Biological basis of breeding and artificial insemination of livestocks* (Russian), Leningrad. 12(2): 154–160.

Bezverkhny, S.A. 1985. *Agricultural Professions of Laser Beam* (Russian), Agropromizdat, Moscow, p 135.

Bezverkhny, S.A. and Pigarev, L.A. 1984. Laser defectoscopy of eggs. *Doklady. VASKHNIL* (Russian) 4: 38–39.

Blackburn, W.E., Bryant, T.D. and Harris, D.I. 1964. Apparatus for detecting blood spots in eggs. Patent USA No 3130255, 21 April 1964.

Bogolubsky, S.I. 1958. Fluorescence of eggs as selection sign. *Proc. of Pushkin Sci. Laboratory of Poultry-Breeding,* Pushkin, USSR, V.8.

Bol, J. 1981. Laser scanning system for the automated inspection of eggs for haircracks. In: *Quality of Eggs, Proc. of 1st European Egg Quality Symposium (Group 4),* G.M. Beuving, C.W. Scheele, P.C.M. Simons, eds. Beekbergen, Netherland, Spelderholt Institute of Poultry Research. p. 84–93.

Brant, A.W. 1953. Machine sorts eggs for shell colour. USDA. *Poultry Processing and Marketing.* 59: 12–13.

Brant, A.W., Norris, K.H. and Chin, G. 1953. A spectrophotometric method for detecting in white-shelled eggs. *Poultry Science* 32: 357–363.

Doran, B.M. and Muellar, W.J. 1961. The development and structure of the vitelline membrane and their relationship to yolk mottling. *Poultry Sci.* 40: 474–478.

Goodrum, J.W. and Elster, R.T. 1992. Machine vision for crack detection in rotating eggs. *Transactions of the American Society of Agricultural Engineers* 35(4): 1323–1328.

Helbaska, A.V.I. and Swanson, M.H. 1958. Studies on blood and meat spots in the hen's egg. 2. Some chemical and histological characteristics of blood and meat spots. *Poultry Sci.* 37: 817–885.

Imai, C. and Saito, J. 1985. Detection of spoiled eggs using a new-type black light egg inspection unit. *Poultry Sci.* 64(10): 1891–1899.

Nalbandow, A.V. and Card, L.E. 1944. The problem of blood clots and meat spots in chicken eggs. *Poultry Sci.* 23: 170–180.

Philip, T., Weber, C.W. and Berry, J.W. 1976. Rapid measurement of egg yolk color. *Food Technol.* 30(11): 58–59.

Polin, D. 1957. Biochemical and weight changes of mottled yolks in eggs from hens fed Nicarbazin. *Poultry Sci.* 36: 831–835.

Posudin, Yu.I. 1988. *Lasers in Agriculture.* Science Publishers, Inc. Enfield, NH, USA. p 188.

Posudin, Yu.I. 1995. *Biophysics.* Urozhaj, Kyiv. p 222.

Posudin, Yu.I. 2005. *Methods of Nondestructive Quality Evaluation of Agricultural and Food Products.* Aristey, Kyiv. p 408.

Posudin, Yu.I. and Kucherov, A.P. 1992. Method of control of eggs quality. Invention N 1783390 USSR, 1992.

Posudin, Yu.I. and Lepeshenkov, V.F. 1991. Optical and laser methods of poultry-breeding control. *St.-Petersburg Agr. Univ. Publ.* 1: 59–64.

Posudin, Yu.I. and Lepeshenkov, V.F. 1992. Method of eggs quality evaluation. Invention N 1735766 USSR, 1992.

Posudin Yu.I., Tsarenko P.P. and Tsyganjuk, O.V. 1992. Eggs quality evaluation with optical and laser methods. *Agr. Biology.* 4: 69–74.

Rybalova, N.B. and Maslova, V.G. 1985. Visual and objective methods of estimation of eggshell luminescence. In: *Development of methods which lead to increasing of poultry production.* Leningrad. p. 22–29.

Solomon, S.E. 1997. *Egg and Eggshell Quality.* Iowa State Press, p 149.

Stadelman, W. J. and Cotterill, O.J. 1973. *Egg Science and Technology.* Westport, Conn, Avi Pub. Co., p 590.

Tzarikov, N.N. and Chernova, G.G. 1968. Application of physical methods in egg product industry. Central Institute of Information, Technical and Economical Investigations. Moscow, pp. 1–18.

Van Deren, J.M. and Jaworski, E.G. 1968. Collaborative study of the determination of exthoxyquin in chick tissue and eggs by fluorescence. *J. of A.O.A.C.* 51: 537-539.

Vikram, C.S., Vedam, K. and Buss, E.G. 1980. Nondestructive evaluation of the strength of eggs by holography. *Poultry Sci.* 59: 2342–2347.

Wells, R.G. and Belyavin, C.G. 1987. Egg Quality – Current Problems and Recent Advances. *Proc. Poultry Science Symp.* London, Boston, Durban, Sydney, Toronto, Wellington. N 20. p 302.

CHAPTER

6

Spectroscopic Analysis of Honey

6.1 PROPERTIES OF HONEY

6.1.1 Definition of Honey

Honey, according to the accepted definition [Codex..., 1969], is the natural sweet substance produced by honey bees from the nectar of plants (*Blossom Honey* or *Nectar Honey*), or from secretions of living parts of plants, or excretions of plant sucking insects on the living parts of plants (*Honeydew Honey*), which the bees collect, transform by combining with specific secretions of their own, deposit, dehydrate, store and leave in the honeycomb to ripen and mature. Honey is a product with complex chemical composition: it contains plant pigments (carotenes, xanthophylls, chlorophyll), mineral substances, sugars, and various impurities.

Physical properties of honey are related to its state, age, presence of water and level of crystallization. All these factors affect the quality of honey.

6.1.2 Composition of Honey

The main components of honey are sugars, which are presented by fructose (37.20%), glucose (31.28%), sucrose (1.31%), maltose (7.31%), etc. [Je'Anne, 1991]. Blossom honey differs from honeydew by the values of simple sugars, disaccharides, higher sugars, acids, mineral salts and nitrogen content. The concentration of sugars can be used as a criterion of honey adulteration, which is provoked with the artificial addition of syrup, sucrose that is hydrolyzed with acids, starch or beetroot treacle in honey [Chudakov, 1967]. That is why precise quality evaluation of honey has long been the goal of many investigators and specialists who are related to honey-breeding [Vorwohl, 1984, 1990; Gonnet, 1986; Vakhonina et al., 1987; Dustmann, 1993; Mautz, 1993; Campos, 1994]. The composition of honey also reflects the contaminants which are present in the area of bee activity.

6.2 SPECTROSCOPIC ANALYSIS OF HONEY

Development of nondestructive methods, which are based on the analysis of the sample without any alterations of product attributes, present the problem of great practical significance. Spectroscopic methods occupy an important place among the comprehensive tests [Posudin, 2005]. These methods include the measurement of difference between input and output light signals during interaction of light with the sample (absorption, transmission, reflection, scattering, re-emission) and analysis of the dependence of this difference on the wavelength. Besides, spectroscopic methods are rather fast and precise. Effects of honey type, age, temperature, water content and degree of sugar adulteration on the spectral properties of honey can also be studied.

Certain spectroscopic methods that have been applied to honey control are mentioned in the literature: spectrophotometry [Yao and Chen, 1985; Yao and Fan, 1985; Salinas et al., 1994], optical activity measurement [Juan et al., 1992], atomic spectroscopy [Petrovic et al., 1994; Salinas et al., 1994]. However, the fact is that honey presents a nontransparent and opaque substance; the application of the abovementioned methods requires either the dilution, or special preparation of the samples. That is why it is necessary to use other spectroscopic methods which are based on the interaction of optical radiation with surface layers of the sample.

Two methods can be chosen from this point of view - fluorescence spectroscopy and reflectance spectroscopy in near-infrared (NIR) part of the spectrum. These methods can effectively provide nondestructive quality evaluation of honey.

The main objective of this investigation is development of the spectrum, based on fluorescence and NIR spectroscopy, approach to nondestructive quality evaluation of honey and establishment of the correlation between those factors which determine honey quality and its spectral properties.

6.3 HONEY SAMPLES

The samples of honeydew, which were used in these experiments, were taken from the collection of Bavarian Institute of Bee-Breeding

(Erlangen, Germany)*, and the samples of honey – from different parts of Ukraine: Kiev, Dnepropetrovsk, and Crimea. Samples with different percentage of sucrose were prepared in the National Agricultural University for obtaining useful information about adulteration of honey. Descriptions of all the samples are given in Table 6.1. The measurements of spectroscopic properties of Bavarian "Tanne" samples were performed in 1993, and of the Ukrainian honey samples in 1994.

Table 6.1 Samples of honey which were used in the experiments

Sl. No	*Type of Honeydew*	*Geographic Origin*	*Year of Collection*
1	Abies alba, Schwarzwald ("Tanne")	Germany, Bavaria	1989
2	"	"	1992
3	Abies alba, Bayerischerwald ("Tanne")	"	1990
	Type of Honey	*Geographic Origin*	*Year of Collection*
4	Monofloral, Acacia	Ukraine, Crimea	1993
5	Monofloral, Sunflower	"	"
6	Polyfloral, Esparcet-Rape-Acacia	"	"
7	Polyfloral, Sage-Lavender	"	"
8	Polyfloral, *Sonchus*-Buckwheat-Sunflower	"	"
9	Acacia	Ukraine, Verkhnedneprovsk	1993
10	"	Ukraine, Pavlovka	"
11	"	Ukraine, Motronovka	"
12	"	Ukraine, Vodyanoye	"
13	"	Ukraine, Dedovo	"
14	"	Ukraine, Andreevka	"
15	"	Ukraine, Maly Bukrin	"
16	"	Ukraine, Kiev	"
17	Lime-Tree	"	1996
18	Multiherbal Collection	"	"

6.4 INSTRUMENTATION

The spectra of absorption and reflection of honey in the ultraviolet and visible part of the spectrum were measured with spectroscopic complex KSVU-23 ("LOMO", Russia), which was equipped with a double monochromator, diffraction gratings, and computer.

The investigation of fluorescence spectra of honey was performed with the spectrofluorometer SDL-2 ("LOMO", Russia) in the regime of

*The samples of honeydew were presented by the courtesy of Dr. D. Mautz.

photon counting. The spectral range during these measurements was 200-700 nm; the errors of measurement were 2 nm for intensity of the bands and 5 nm for half-width of the bands.

The reflectance spectra in the near-infrared (NIR) part of spectrum (1,620-2,320 nm) were measured with analyzer Model 4250 ("Pacific Scientific", USA). This analyzer has three ranges: 1,620-1,800 nm, 1,890-2,115 nm, and 2,050-2,320 nm. Reproducibility of the results was better than 0.015 nm. The NIR spectrum was estimated as the dependence of optical density $D = \lg(1/R)$ on the wavelength λ (where R is reflection coefficient). All measurements were performed at room temperature.

6.5 RESULTS OF SPECTROSCOPIC ANALYSIS OF HONEY

The *absorption spectra* of honeydew ("Tanne") in ultraviolet and visible parts of the spectrum are presented in Fig. 6.1. A certain shoulder near 250-275 nm is standing out against the background which is decreasing monotonously from 200 to 700 nm. The presence of the shoulder in absorption (reflectance) spectra of honey testifies the participation of several honey components in formation of absorption (reflectance) spectra in the ultraviolet and visible parts of the spectrum. The intensity of absorption can be used as a criterion of geographic origin or age of honey.

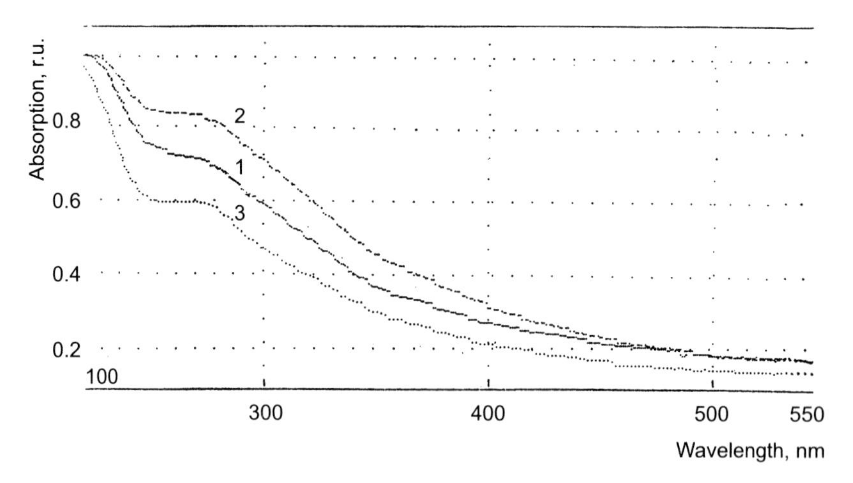

Fig. 6.1 Absorption spectra of honey in ultraviolet and visible parts of spectrum. 1 – Schwarzwald "Tanne", 1989; 2 – Schwarzwald "Tanne", 1992; 3 – Bayerischer "Tanne", 1990 [Posudin et al., 1995]

The *excitation fluorescence spectra* of honeydew were investigated for different wavelengths of fluorescence emission (440 nm, 560 nm, and 575 nm).

The *emission fluorescence spectra* of honeydew are characterized by a broad band (about 100–150 nm); the spectral position of its maximum depends on the wavelength of excitation. These maxima are located near 420 nm (λ_{exc} = 350 nm), 480 nm (λ_{exc} = 400 nm), and 510 nm (λ_{exc} = 450 nm). The maximal intensity of emission takes place during excitation in the ultraviolet part of the spectrum. The fluorescence intensity of honeydew depends on the geographical origin and its age. Typical fluorescence emission spectra of honeydew are presented in Fig. 6.2.

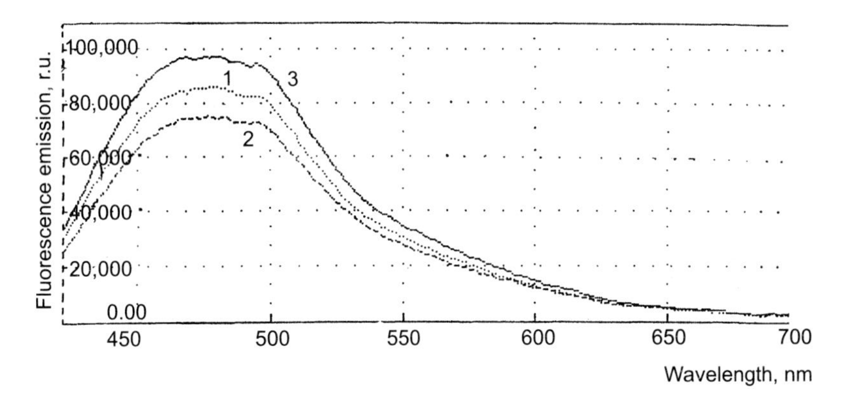

Fig. 6.2 Fluorescence emission spectra of honey (samples N1-N3). Wavelength of excitation 400 nm [Posudin et al., 1995]

The fluorescence intensity of honey also depends on the type of honey (Table 6.2); the samples of the same type of honey (Acacia), which were collected from different parts of one region and of one age, demonstrated quasi equal intensity of fluorescence, but this intensity depends on the age of honey (Table 6.3).

Table 6.2 Dependence of fluorescence intensity of honey on the geographic origin

SL. No. of Sample (From Table 6.1)	*Type of Honey*	*Fluorescence Intensity*
4	Monofloral, Acacia	0.29 ± 0.064
5	Monofloral, Sunflower	0.48 ± 0.088
6	Polyfloral, Esparcet-Rape-Acacia	0.38 ± 0.031
7	Polyfloral, Sage-Lavender	0.69 ± 0.096
8	Polyfloral, Sonchus-Buckwheat-Sunflower	0.43 ± 0.045

Table 6.3 Dependence of fluorescence intensity on the geographic origin and year of collection of honey

SL. No. of Sample (From Table 6.1)	*Geographic Origin*	*Year 1993*	*Year 1994*
9	Verkhnedneprovsk	0.30 ± 0.08	0.53 ± 0.07
10	Pavlovka	0.21 ± 0.05	0.68 ± 0.07
12	Vodyanoye	0.31 ± 0.09	0.82 ± 0.16
13	Dedovo	0.34 ± 0.10	0.44 ± 0.03
14	Andreevka	0.28 ± 0.09	0.50 ± 0.02
15	Maly Bukrin	1.28 ± 0.06	2.05 ± 0.09

The fact of the dependence of intensity, half-width, and spectral position of fluorescence spectra on the wavelength of excitation and emission means the participation of several fluorophors in formation of fluorescence spectra of honey.

The effect of temperature on the fluorescence intensity of honey studied is demonstrated in Fig. 6.3. It is shown that the increasing of temperature provokes the decreasing of the fluorescence intensity.

The correlation between fluorescence properties of honey and the presence of water in it was also investigated. The quantity W of water was estimated with the refractometric method according to the following relation [Aganin, 1989]:

$$W = 400[1.538 - n(20°\text{C})] \quad \text{(Eqn. 6.1)}$$

where n is the coefficient of refraction. The results of these measurements are presented in Table 6.4.

The coefficients of correlation were calculated for polyfloral honey ($r = -0.96$) and monofloral honey ($r = -0.87$). This strong negative correlation between the presence of water in honey and fluorescence intensity can be used for quality evaluation of honey.

The *reflectance spectra* of honey samples in NIR part of the spectrum are characterized by a number of reflectance bands near 1,779 nm, 1,933 nm, and 2,290 nm; the relative intensity of the spectral bands depends on the type and age of the sample (Fig. 6.4). The samples taken from different parts of the same geographic region produced the same shape of the reflectance spectra, which were distinguished by the intensity of spectral bands only (Figs. 6.5 and 6.6). However, the samples that were taken from different geographic zones demonstrated different shapes (Figs. 6.4 and 6.5).

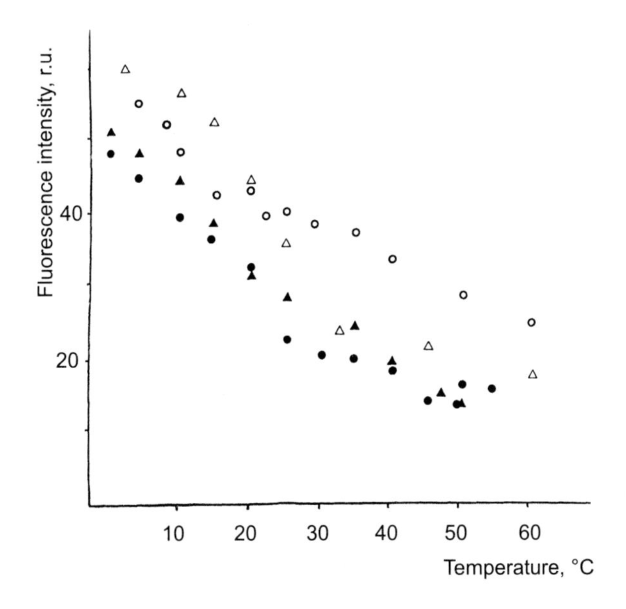

Fig. 6.3 Dependence of fluorescence intensity of honey on the temperature. -O- – monofloral, acacia, sample N4; -▲- – monofloral, sunflower, sample N5; -Δ- – polyfloral, sage-lavender, sample N7; -•- – polyfloral, sonchus-buckwheat-sunflower, sample N8 [Posudin et al., 1995]

Table 6.4 Results of estimation of quantity *W* of water by fluorescence and refractometric methods

Index of Refraction (N)	*Quantity of Water (W)*	*Fluorescence Intensity (I)*	*Index of Refraction (n)*	*Quantity of Water (W)*	*Fluorescence Intensity (I)*
	Sage-Lavender, Sample N7			*Acacia, Sample N4*	
1.4961	16.76	2.98	1.4976	16.16	1.45
1.4835	21.80	2.22	1.4785	23.80	1.28
1.4711	26.80	1.52	1.4651	29.20	0.96
1.4651	29.20	1.34	1.4565	32.60	0.88
1.4420	37.50	0.92	1.4282	44.00	0.70
1.4222	46.40	0.78	1.4082	51.90	0.64
1.4061	52.80	0.74	1.3871	60.40	0.62
1.3849	61.20	0.66	1.3600	71.20	0.50
1.3650	68.20	0 56	1.3500	75.20	0.32
1.3489	75.60	0.42	1.3380	80.00	0.24
1.3380	80.00	0.34	–	–	–

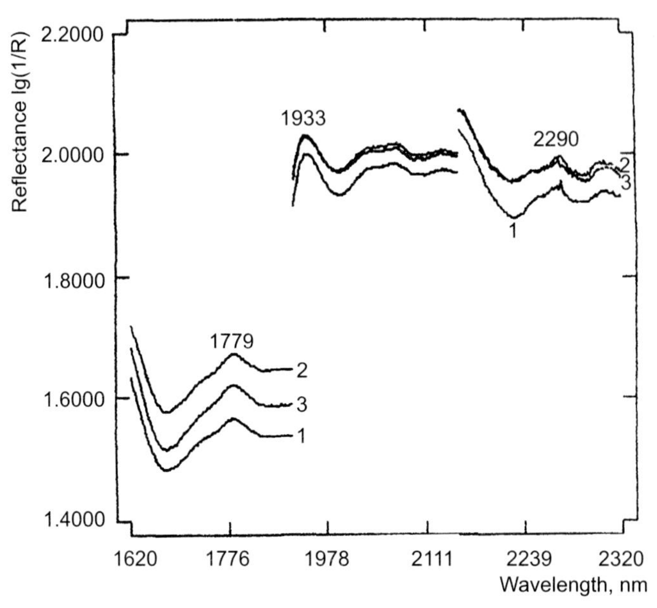

Fig. 6.4 Reflectance spectra of honey in near-infrared part of spectrum (samples N1-3) [Posudin et al., 1995]

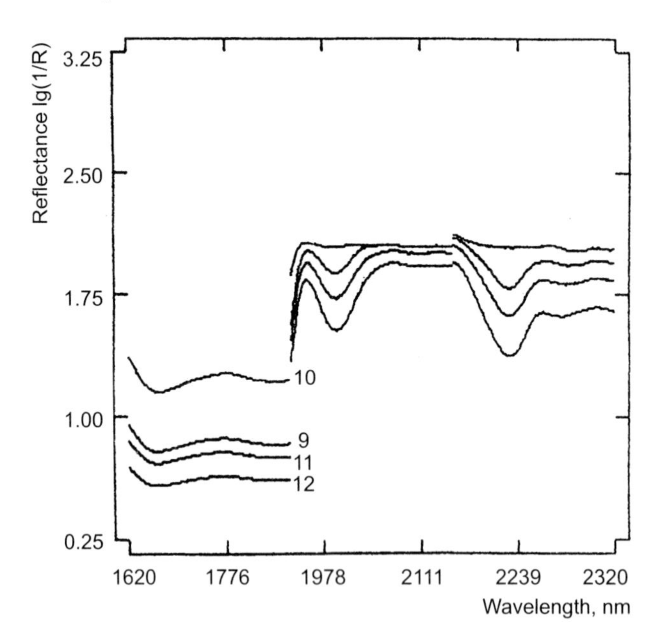

Fig. 6.5 Reflectance spectra of honey in near-infrared part of spectrum (samples N9-12) [Posudin et al., 1995]

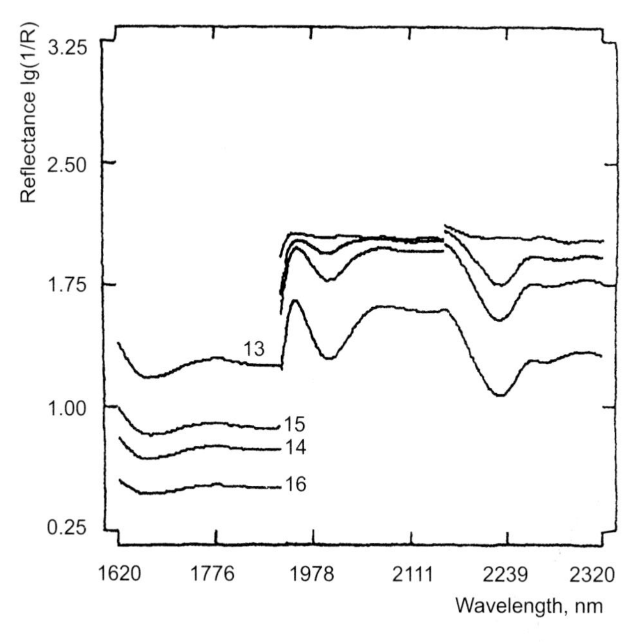

Fig. 6.6 Reflectance spectra of honey in near-infrared part of the spectrum (samples N13-16) [Posudin et al., 1995]

It is very informative to use spectral parameters of honey for the nondestructive detection of honey adulteration. Figures 6.7 and 6.8 demonstrate the evolution of the reflectance spectra for the samples with different concentration of sucrose, which was artificially added to honey from lime-tree and multiherbal collection respectively. The curves "reflectance versus sucrose concentration" which are presented in Fig. 6.9, can be used for quantitative estimation of honey adulteration. The accuracy limits in these experiments did not exceed 5%.

Thus, honey is characterized by certain spectral properties in ultraviolet, visible, and near-infrared parts of the spectrum, which are related to its ability to absorb, transmit, reflect and re-emit optical radiation. The chemical composition and physical properties of honey are closely related to its spectral parameters, which can be used as taxonomic indices or indicators of honey state and quality.

6.6 SUMMARY

Spectroscopic methods can be explored in honey-breeding. All the types of interaction of optical radiation with honey can be used in principle for fast and precise diagnostics of this product. The chemical composition,

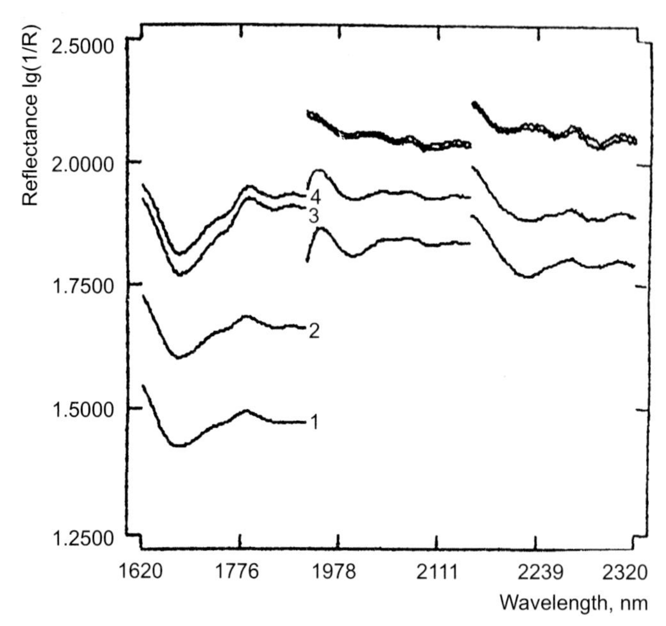

Fig. 6.7 Dependence of reflectance spectra of lime-tree honey (sample N17) in near-infrared part of spectrum on the concentration of sucrose which was artificially added to honey. 1 – natural honey; 2 – 20% sucrose; 3 – 40% sucrose; 4 – 60% sucrose [Posudin et al., 1995]

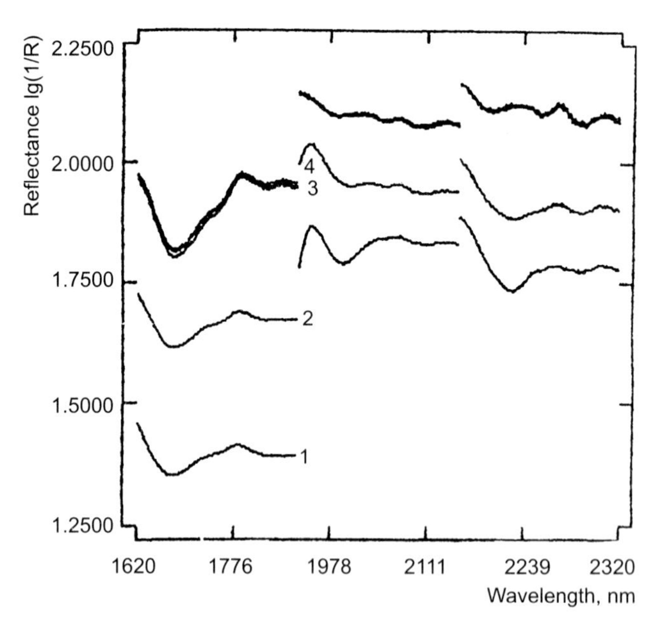

Fig. 6.8 Dependence of reflectance spectra of multiherbal honey in near-infrared part of spectrum on the concentration of sucrose which was artificially added to honey. 1 – natural honey; 2 – 20% sucrose; 3 – 40% sucrose; 4 – 60% sucrose [Posudin et al., 1995]

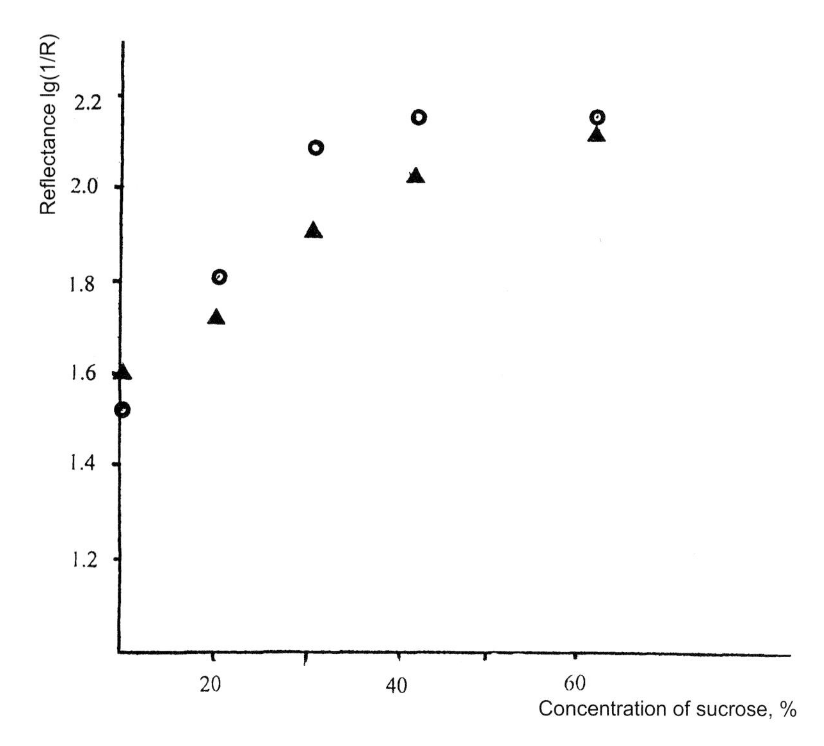

Fig. 6.9 Dependence of reflectance of honey on the concentration of sucrose. Wavelength is 1,779 nm. -▲- – lime-tree honey, -o- – multiherbal honey [Posudin et al., 1995]

physical properties, geographic origin and age of honey are closely related to its spectral parameters which can be used as taxonomic indices or indicators of honey state and quality.

The near-infrared reflectance spectroscopy gives useful information about the adulteration of honey.

6.7 REFERENCES

Aganin, A.V. 1989. *Sanitary Expertise of Honey*. Saratov Agr. Inst. Publ., Saratov, Russia. p 40.

Campos, M.G. 1994. Controle de qualidade e tipificacao de meis, *Apicultor*. 2(6): 19–24.

Chudakov, V.G. 1967. *Composition and Properties of Sugar Honey and Methodics of Detection of this Adulteration*. Moscovsky Rabochy, Moscow, p 130.

Codex Alimentarius Comission CAS/RvS 12/1969, *Recommended European Regional Standard for Honey*, FAO/WHCXed., Rome. Codex Stan 12–1981, Rev.1 (1987), Rev.2 (2001).

Dustmann, J.H. 1993. Honey, quality and its control. *American Bee Journal*, 133(9): 648–651.

Gonnet, M. 1986. L'analyse des miels. Description de queiques methodes de controle de la qualite, *Bulletin-Technique-Apicole,* 13(1): 17–36.

Jeanne, F. 1991. Le Miel. Definition, origines composition et proprietes.

Je'Anne, F. 1991. Le Miel. Definition, origines composition et proprietes, *Buttetin-Technique-Apicole,* 18(3), 76: 205–210.

Juan, T., Conchello, M.P., Tello, M.L., Perez-Arquille, C and Herrera, A. 1992. Rotation especifica y espcetro glucidico de mieles de Zaragoza, *Attmentaria,* 28(232): 75–78.

Mautz, D., Rosenkranz, P. and Schaper, F. 1993. Die Tätigkeit der B. Landesanstalt für Bienenzucht. Erlangen im Jahre 1991/92, *Imkerfreund,* 9: 1–29.

Petrovic, Z.T., Mandic, M.I., Grgic, J. and Grgic, Z. 1994. Ash and chromium levels of some types of honey, *Zeitschrift für Lebensmittel-Untersuchung und Forschung,* 198(1): 36-39.

Posudin, Yu.I. 1995. *Biophysics.* Urozhaj, Kyiv. p 222.

Posudin, Yu.I. 2005. *Methods of Nondestructive Quality Evaluation of Agricultural and Food Products.* Aristey, Kyiv. p 408.

Posudin, Yu.I., Polishuk V.P., Bulavin S.P. and Istomina, V.A. 1995. Spectroscopic methods of honey investigation. *Visnyk agrarnoi nauki.* 6: 78–82.

Salinas, F., Deespinosa, V.M., Osorio, E. and Lozano, M. 1994. Determination of mineral elements in honey from different floral origins by flow-injection analysis coupled to atomic spectroscopy, *Revista Espanola De Ciencia Y Technologia De Ailmentos,* 34(4): 441–449.

Salinas, F., Espinosa-Mansilla, A. and Berzas-Nevado, J.J. 1994. Simultaneous determination of sulfathiazole and oxytetracycline in honey by derivative spectrophotometry, *Microchemical Journal,* 43(3): 244-252.

Vakhonina, T.V., Levina, L.P. and Bondareva, K.M. 1987. Quality of honey-pollen products, *Pchelovodstvo.* Moskva: Agropromizdat, 8: 28–29.

Vorwohl, G. 1990. Honigquatitatskontrolle und Honigqualitatskriterien in der Bundesrepublik, *Bienenvater,* 111(11): 391–398.

Vorwohl, G. 1984. Honey and other hive products, and their quality control, *Proc. of the Third Intern. Conf. on Apiculture in Tropical Climates,* Nairobi, Kenya, 5–9 Nov. 1984, p. 169–170.

Yao, C. and Chen, L. 1985. Determination of traces of lead in honey by spectrophotometry with tetra(p-trimethylammoniumphenyl)porphine, *Shipin-Yu-Fajiqo-Gongue,* 6: 38–41.

Yao, C. and Fan, C. 1985. Determination of zinc in honey by spectrophotometry with 5, 10, 15, 20-tetrakis (4-trimethylammoniophenyl) porphine, *Shipin-Yu-Fajiao-Gongue,* 2: 17–20.

CHAPTER

7

Spectroscopic Analysis of Animal Hair and Bird Feathering

7.1 PROPERTIES OF ANIMAL HAIR-COVERING

7.1.1 Structure of Hair-covering

A single hair consists of root with follicle, and hair shaft, which has components such as core, scab layer and cuticle (Fig. 7.1). The hair-covering of the farm animals is characterized by the following parameters: the density of the hairs (1 cm^2) is: 2,600 (cow), 3,000–4,000 (sheep), 700 (horse); the average thickness of a hair is (in μm): 20–100

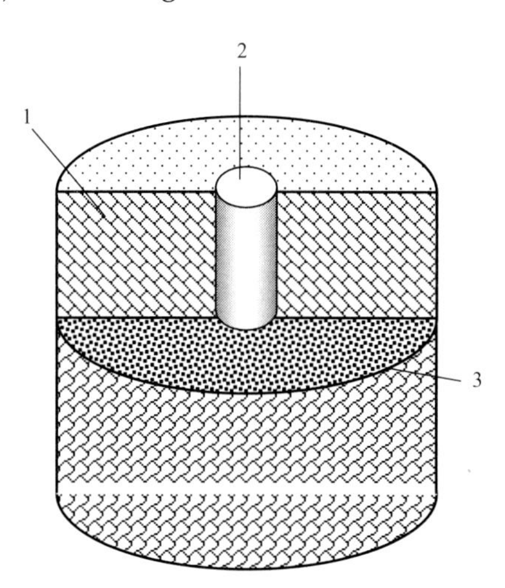

Fig. 7.1 A single hair consists of root with follicle, and hair shaft, which has components such as: 1 – scab layer, 2 – core, and 3 – cuticle

(cow), 5–40 (sheep), 10–200 (horse). The density of hair is about 700 cm^{-2} for horse, 5,000–6,000 cm^{-2} for sheep.

7.1.2 Colouration of Animal Hair-covering

The colouration of animal hair-covering is determined by quantitative and qualitative characteristics of hair pigments. *Melanin* is the brown pigment which is produced by the specialized cells – melanocytes. These cells are located in the hair core. It is possible to distinguish three main groups of melanines – *eumelanines*, that are responsible for black and brown colour; *pheomelanines*, that produce yellow and red colour; *trichochroms* with low molecular weight. The dimensions of melanine granules in the hair are about 0.5–1.2 µm for dark colour hair and 0.1–0.3 µm for white hair. The relative concentration of melanine depends on the colour of the hair cover and consists of 8.2–8.4 for dark and about 1.7 for light colour (% from mass of the sample).

The structure, composition and properties of pigments depend on the phase of hair growth [Seiji, 1973], content or deficiency of metals [Kostaneski, 1970], presence of vitamins *A* and *D* in feeding [Wojcikowska-Soroczynska et al., 1987], solar radiance [Guidance to study skin-covering of mammals, 1988], and violation of blood vessel system [Durst, 1963].

As a whole, colour of the hair can be considered as an important genotypical sign [Ioganson et al., 1970].

7.2 HAIR SAMPLES

The first group of the samples was presented by the hair-covering of the horses from sportive club of National Agricultural University. Different types of colouration such as black, bay, grey and chestnut were chosen. The samples were taken from the head, side and mane of the horse [Posudin et al., 1993].

The second group of the samples were taken from the hair-covering of ponies. The samples were chosen from the head, neck, side, croup, extremities and nails of the horses [Posudin et al., 1993].

The third group of the samples used were from the hair-covering of the Przevalsky horse (*Equus przewalskii*) taken from the collection of Ukrainian Research Institute of the Cattle-Breeding of Steppe Regions "Askaniya-Nova". The samples were taken from the cheek, neck, shoulder blade, back, side, stomach, croup, extremities, mane, and tail (Fig. 7.2) of 22 different horses [Posudin et al., 1998]. Two main colour types of the Przevalsky horse within the same herd are known [Poliakov,

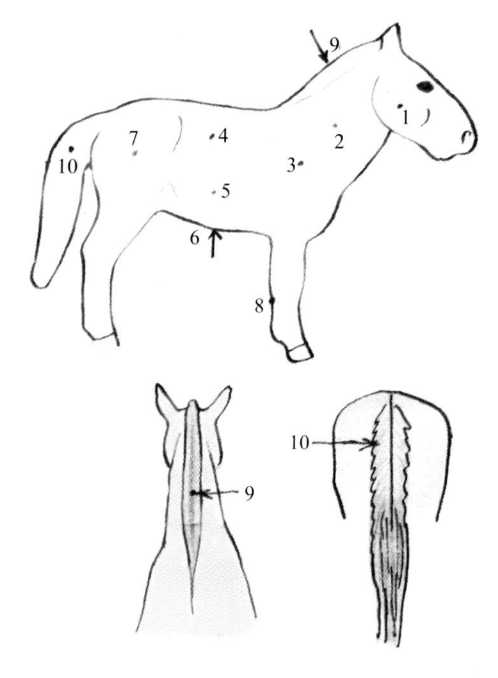

Fig. 7.2 Samples taken from 22 different horses. 1 – cheek, 2 – neck, 3 – shoulder-blade, 4 – back, 5 – side, 6 – stomach, 7 – croup, 8 – extremities, 9 – mane, 10 – tail [Posudin et al., 1998]

1881; Bannikov, 1958; Groves, 1986; Klimov, 1990]: one pale grey-yellow, the other a bright yellowish red-brown. There are intermediate types of colouration also. A dark mono-coloured type can be distinguished which has less marked contrast between the dark head neck and the body; this type has dark muzzle and black legs (Fig. 7.3); a greyish type with the head and neck that are darker than the body; this type has a white muzzle, grey hindlegs and dark grey forelegs, and several thin dark zebra-like stripes on the carpus (Fig. 7.4); the others are light types, either normal with a white muzzle, zebra-like stripes on the carpus, light grey forelegs and light hindlegs (Fig. 7.5), or mono-coloured with a white muzzle, light forelegs and very light hindlegs (Fig. 7.6).

The fourth group of the animal hair that was used in the experiments with laser diffractometry consisted of the vibrissa of 371 heads of bull calves and heifers which were compared with the mass of the animals [Posudin and Trofimenko, 1994].

Fig. 7.3 A dark mono-coloured type which has less marked contrast between the dark head, neck and the body; this type has a dark muzzle and black legs [Posudin et al., 1998]

Fig. 7.4 A greyish type with the head and neck that are darker than the body; this type has a white muzzle, grey hindlegs and dark grey forelegs, and several thin dark zebra-like stripes on the carpus [Posudin et al., 1998]

7.3 INSTRUMENTATION

7.3.1 Spectrometer

The spectrometer which was used for microphotometry and microfluorometry of animal hair consisted of incandescent lamp OI-9 *1*, lens 2, sample *3*, microscope LUMAM-3M *4*, photomultiplier FEU-79 *5*, recorder H-306 *6*, digital voltmeter V7-21 *7*, mercury lamp DRSh-250 *8*,

Fig. 7.5 A light normal type with a white muzzle, zebra-like stripes on the carpus, light grey forelegs and light hindlegs [Posudin et al., 1998]

Fig. 7.6 A mono-coloured light type with a white muzzle, light forelegs and very light hindlegs [Posudin et al., 1998]

glass infrared filter *9*, water infrared filter *10*, halogen lamp KGM9-70 *11*, monochromator UM-2 *12*, and semitransparent plate *13* (Fig. 7.7).

The procedure of microphotometry included the passage of light from the incandescent lamp through the hair and microscope and subsequent registration of the transmitted light by the photomultiplier. A signal $S = (I - I_N)/(I_0 - I_N)$ was registered during microphotometry of the hair, where I and I_0 are signals which were recorded with and without the hair correspondingly, and I_N was the level of noise which was estimated with the closed photomultiplier.

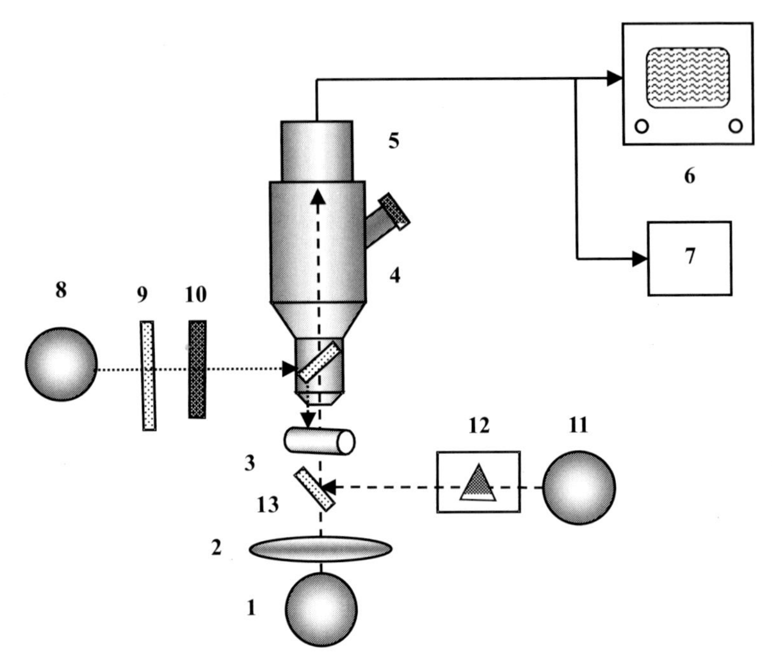

Fig. 7.7 The spectrometer which was used for microphotometry and microfluorometry of animal hair: 1 – incandescent lamp OI-9; 2 – lens; 3 – sample; 4 – microscope LUMAM-3M; 5 – photomultiplier FEU-79; 6 – recorder H-306; 7 – digital voltmeter V7-21; 8 – mercury lamp DRSh-250; 9 – glass infrared filter; 10 – water infrared filter; 11 – halogen lamp KGM9-70; 12 – monochromator UM-2; 13 – semitransparent plate [Posudin et al., 1991]

Microfluorometry of hair was based on the excitation of fluorescence of the sample by radiation of the mercury lamp which passed through filters on the sample. Fluorescence emission radiation was registered by the photomultiplier and readout system.

The measurement of the illuminance of the sample during the experiments was realized by luxmeter Yu-116.

7.3.2 Laser Diffractometry

Let us consider the situation when laser radiation is made to pass through a single slit. The author and his colleagues used He:Ne-laser LGN–215 (wavelength 632 nm) in their experiments. Some of the radiation appears in the geometric shadow of the aperture. It is shown [Born and Wolf, 1970] that the intensity of monochromatic light diffracted by a slit is related to the width d of the slit, the angle of diffraction θ and the wavelength λ of light as follows:

$$I = I_0 \sin \alpha / \alpha^2 \qquad \text{(Eqn. 7.1)}$$

where I_0 is the intensity of the incident light and $\alpha = d \sin \theta/\lambda$.

It is possible to show with certain accuracy that the intensity of light diffracted by a thin cylinder (animal hair or vibrissa) of the same size d can be described by the following equation:

$$I = I_0 (1 - \sin \alpha / \alpha)^2 \qquad \text{(Eqn. 7.2)}$$

While measuring the interval between diffraction maxima of a certain order of diffraction, it is possible to estimate the diameter d of the slit or cylinder (the thickness of the hair). This is the main idea of laser diffractometry. This interval l_m between diffractional maxima and distance L between the sample and the screen where diffraction is observed are related to the expression:

$$\sin \theta_m = \frac{l_m}{\sqrt{l_m^2 + L^2}} \qquad \text{(Eqn. 7.3)}$$

where θ_m is the angle of registering the diffraction maximum of the m-th order (Fig. 7.8).

As soon as the condition of observation of diffraction maxima can be written as:

$$\alpha_m = \frac{4m-1}{2}\pi = \frac{\pi d \sin \theta_m}{\lambda} \qquad \text{(Eqn. 7.4)}$$

it is possible to obtain the expression which connects the diameter d of the hair with parameters l_m and L:

$$\alpha_m = \frac{4m-1}{2}\lambda \sqrt{\frac{l_m^2 + L^2}{l_m}} \qquad \text{(Eqn. 7.5)}$$

7.4 RESULTS OF SPECTROSCOPIC ANALYSIS OF ANIMAL HAIR

7.4.1 Microphotometry and Microspectrophotometry of Horse Hair

The colour of horses is a rather important biological sign; each horse race has a dominating colour. It is known [Combs, 1987] that the colour of the horse hair is a specific indicator of the accumulation of calcium and lead. The colour of the hair also determines the ability of the horse to resist the

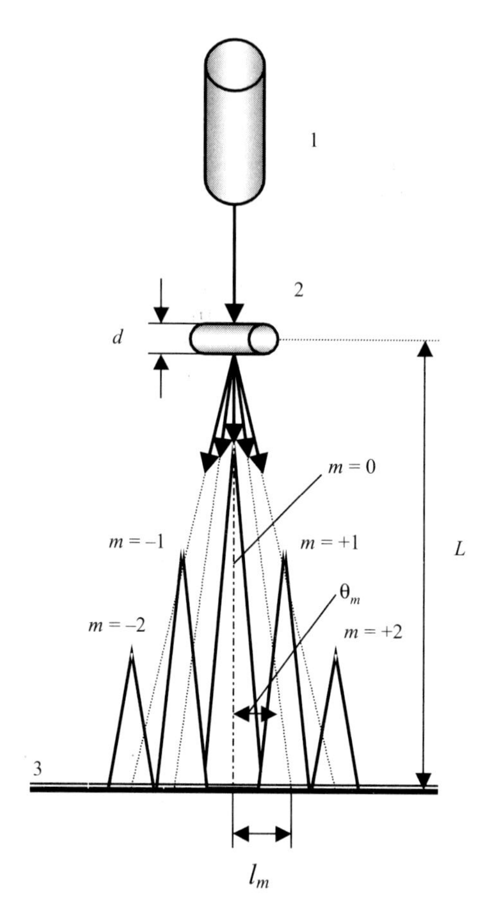

Fig. 7.8 Principle of laser diffractometry: l_m – interval between diffractional maxima; L – distance between the sample and the screen where diffraction is observed; θ_m – the angle of registering the diffraction maximum of the *m*-th order [Posudin et al., 1991]

photosensitizing action of fagopyrin which is found in *Fagopyrum (Polygonum* or *F. esculentum)* during solar irradiation. For example, grey horses suffer more form fagopyrism which leads to melanosarcoma [Gopka et al., 1989]. In addition, the colour of horse hair is related to the relative ratio of melanine and 3,4-dioxyphenylalanine [Durst, 1936].

The dependence of transmittance τ of horse hair on the illuminance E is presented in Figs. 7.9–7.12. It is clear that the curves $\tau = f(E)$ depend strongly on the colour and the part of animal body. Each curve is characterized by a linear part of dependence $\tau = f(E)$ and the region of saturation, where $\tau \rightarrow 1$. The hair of grey colouration has maximal

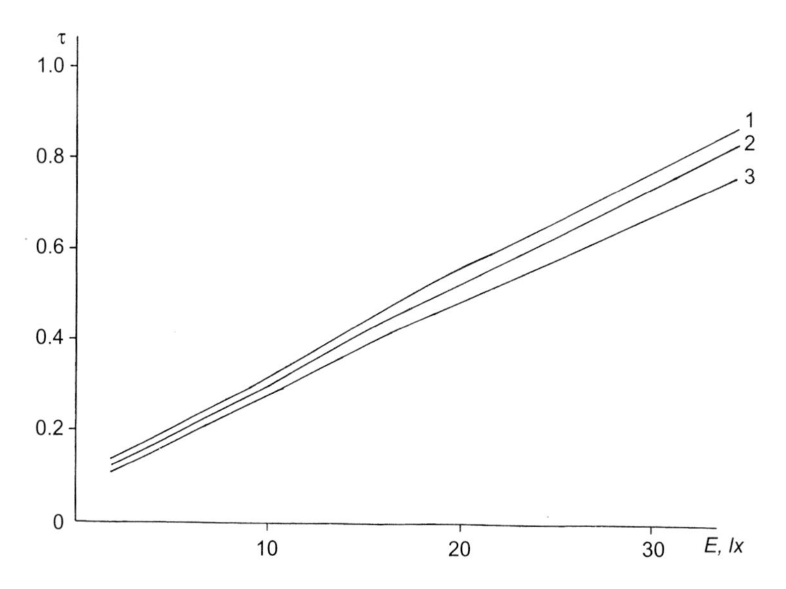

Fig. 7.9 Dependence of transmittance τ of black horse hair on the illuminance *E*. 1 – side; 2 – mane; 3 – head [Posudin et al., 1991]

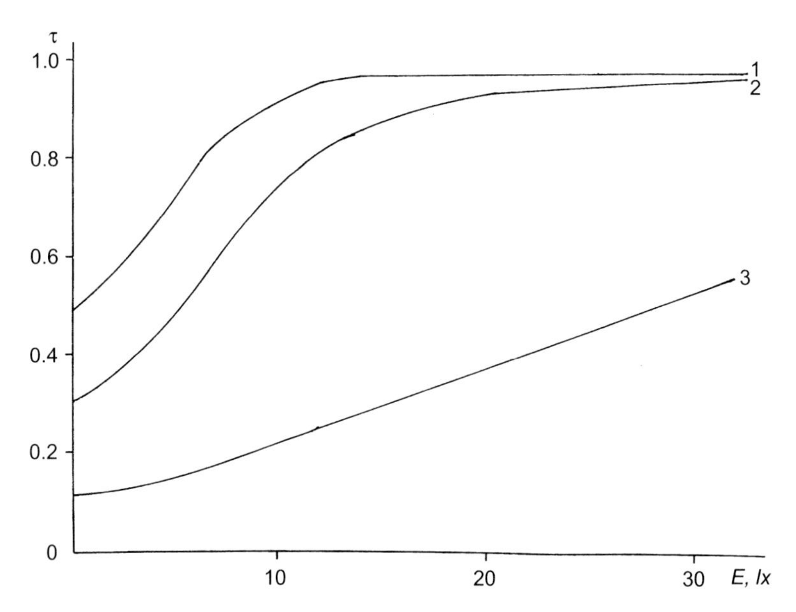

Fig. 7.10 Dependence of transmittance τ of bay horse hair on the illuminance *E*. 1 – head; 2 – side, 3 – mane [Posudin et al., 1991]

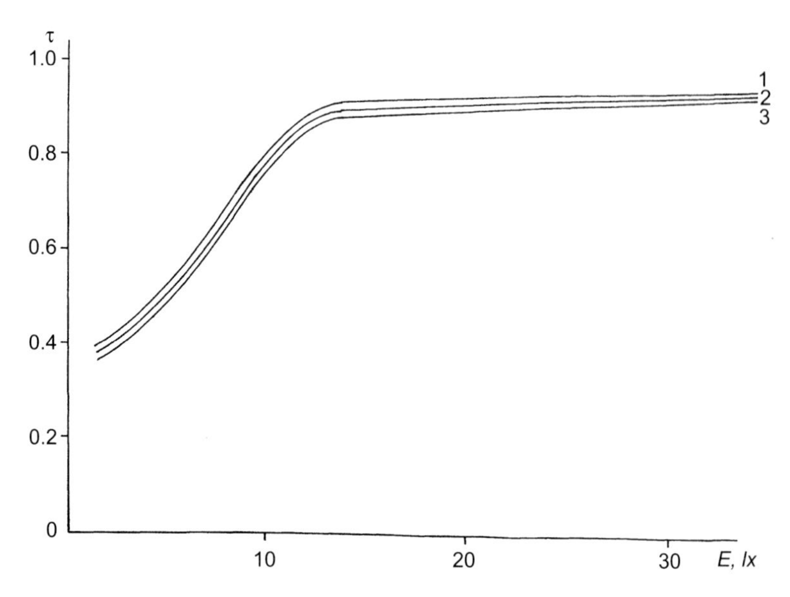

Fig. 7.11 Dependence of transmittance τ of chestnut horse hair on the illuminance *E*. 1 – head; 2 – side; 3 – mane [Posudin et al., 1991]

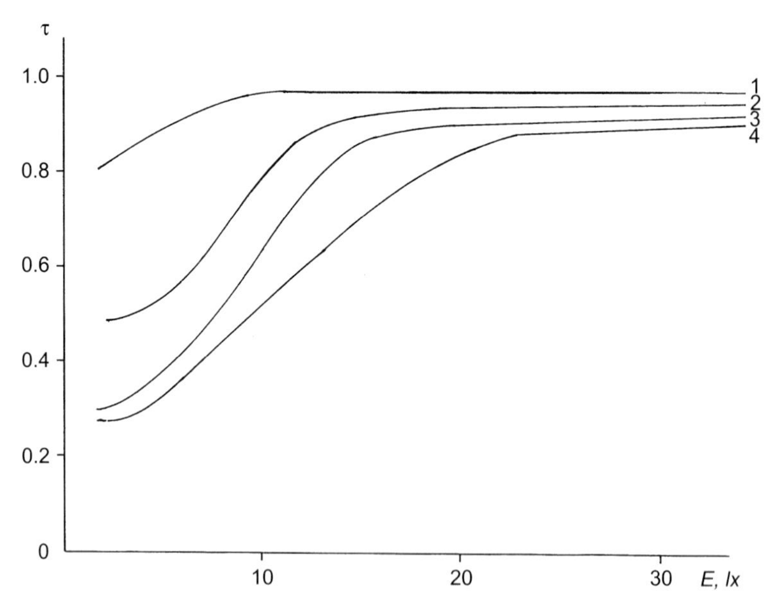

Fig. 7.12 Dependence of transmittance τ of grey horse hair on the illuminance *E*. 1 – mane (core), 2 – mane (crust), 3 – head, 4 – side [Posudin et al., 1991]

transmission, while minimal transmission is typical for the hair of black colouration. The high values of light transmission by grey hair can be explained by early age grizzling, stopping formation of pigments and decreasing of the alkalinity of blood [Durst, 1936]. The linear dependence $\tau = f(E)$ can be used for quantitative estimation of colour of hair-covering of horses.

The typical spectra of transmission of the horse hair are presented in Figs. 7.13–7.16. The character of these spectra depends on the colour, part of the hair (peripherical or central) and the region of animal body. The microphotometry of the hair through the core demonstrates maximal transmission in long-wavelength region for black, chestnut and grey colour. The hair of each colour has different half-widths of transmission bands (black and chestnut colours have maximal half-widths). Transmission of light in the short-wavelength region makes up about 30–40%. The character of transmission spectrum of horse hair depends on the pigment content, the process of light refraction inside the hair, illuminance conditions and distribution of air in internal cavities of the hair [Durst, 1936]. The transmission spectra which were registered by the author can be used for quantitative and qualitative estimation of pigment content in the hair.

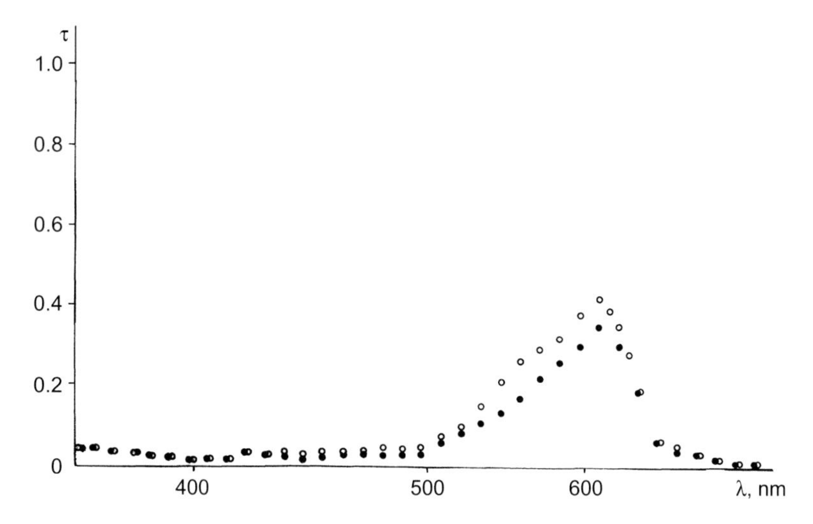

Fig. 7.13 Transmittance spectra of the hair for black colour horse: - o - – head (core); - • - – head (crust) [Posudin et al., 1991]

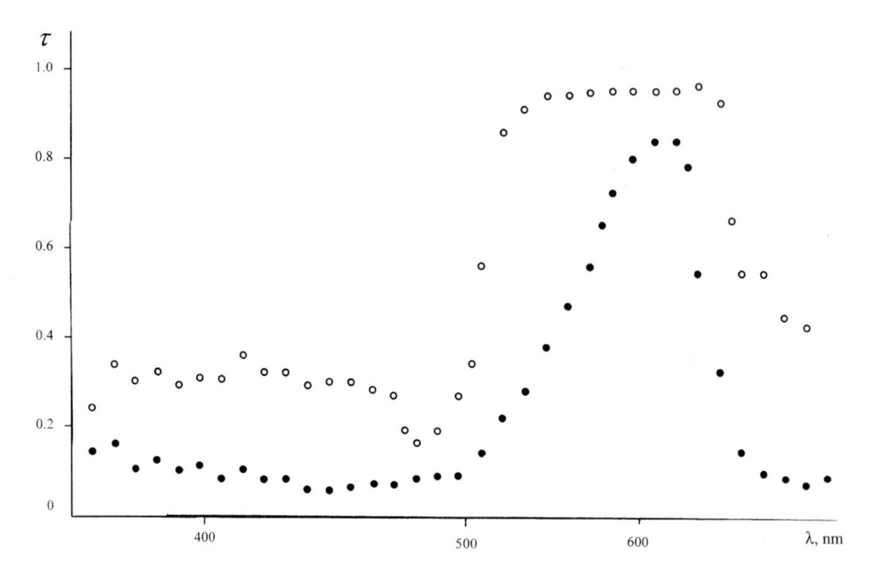

Fig. 7.14 Transmittance spectra of the hair for chestnut colour horse: - o - – head (core); - ● - – head (crust) [Posudin et al., 1991]

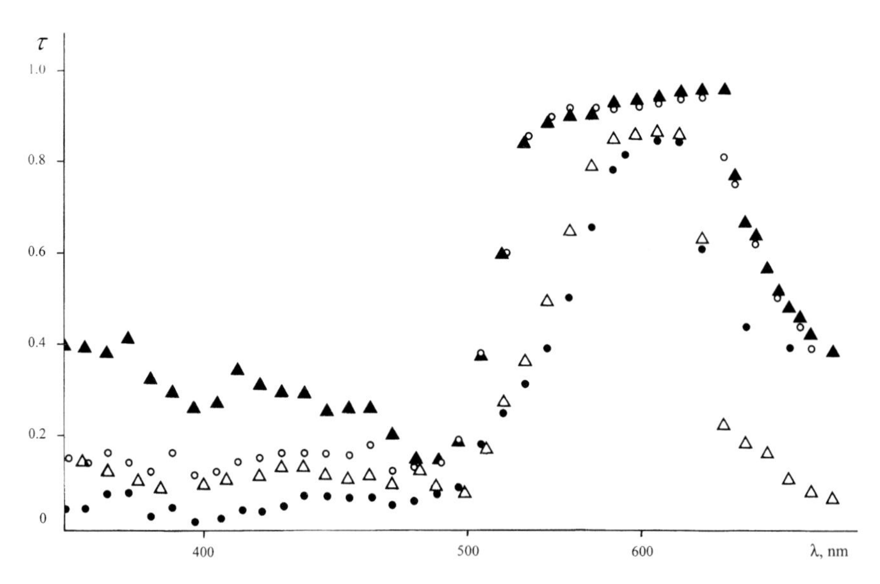

Fig. 7.15 Transmittance spectra of the hair for bay colour horse: - Δ - – head (crust); - ▲ - – head (core); - o - – neck (core); - ● - – neck (crust) [Posudin et al., 1991]

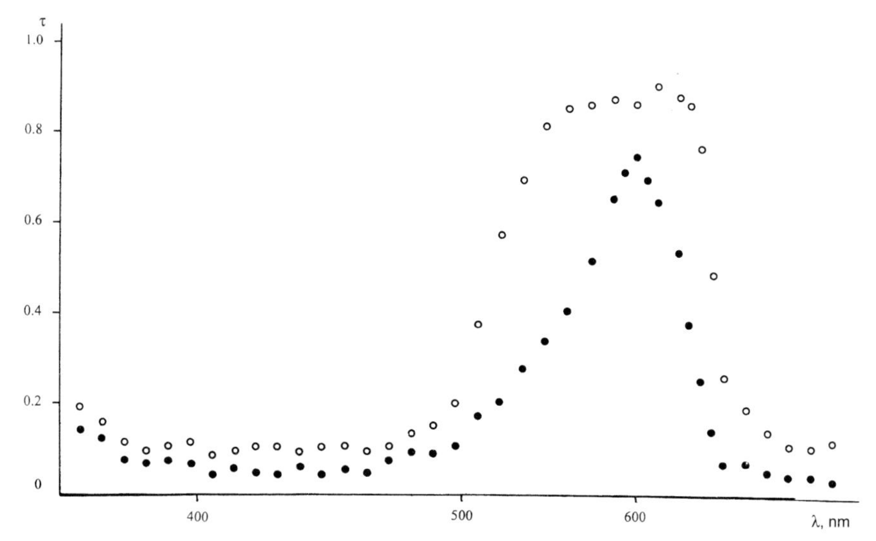

Fig. 7.16 Transmittance spectra of the hair for grey colour horse: - o - – head (core); - • - – head (crust) [Posudin et al., 1991]

7.4.2 Microfluorometry of Horse Hair

Effect of parts of the body. The method of microfluorometry was applied to the quantitative estimation of pigmentation of the hair-covering of sportive horses, ponies and the Przevalsky horse [Posudin et al., 1991; 1993; 1998].

Dependence of the fluorescence intensity of the hair on the part and colour of the sportive horse body is presented in Fig. 7.17. The maximal intensity is observed for the protective hair of light grey horses. If the colour of grey hair can be taken into account as 100%, the colour of the integumentary hair makes up 60% for light grey colour, 30% for dark grey colour, 13–18% for black colour, 25–30% for bay colour and 27–37% for chestnut colour.

Dependence of the fluorescence intensity of the hair on the part of pony body (stallion of black colouration, mare of chestnut colouration and their foal) is presented in Fig. 7.18.

The intensity of fluorescence of the hair samples of the Przevalsky horse depends on the part of horse body (22 individuals have been investigated). Such parts as back, side, croup and shoulder blade demonstrate maximal values of fluorescence, while tail, extremities and mane – minimal values (Table 7.1, Fig. 7.19).

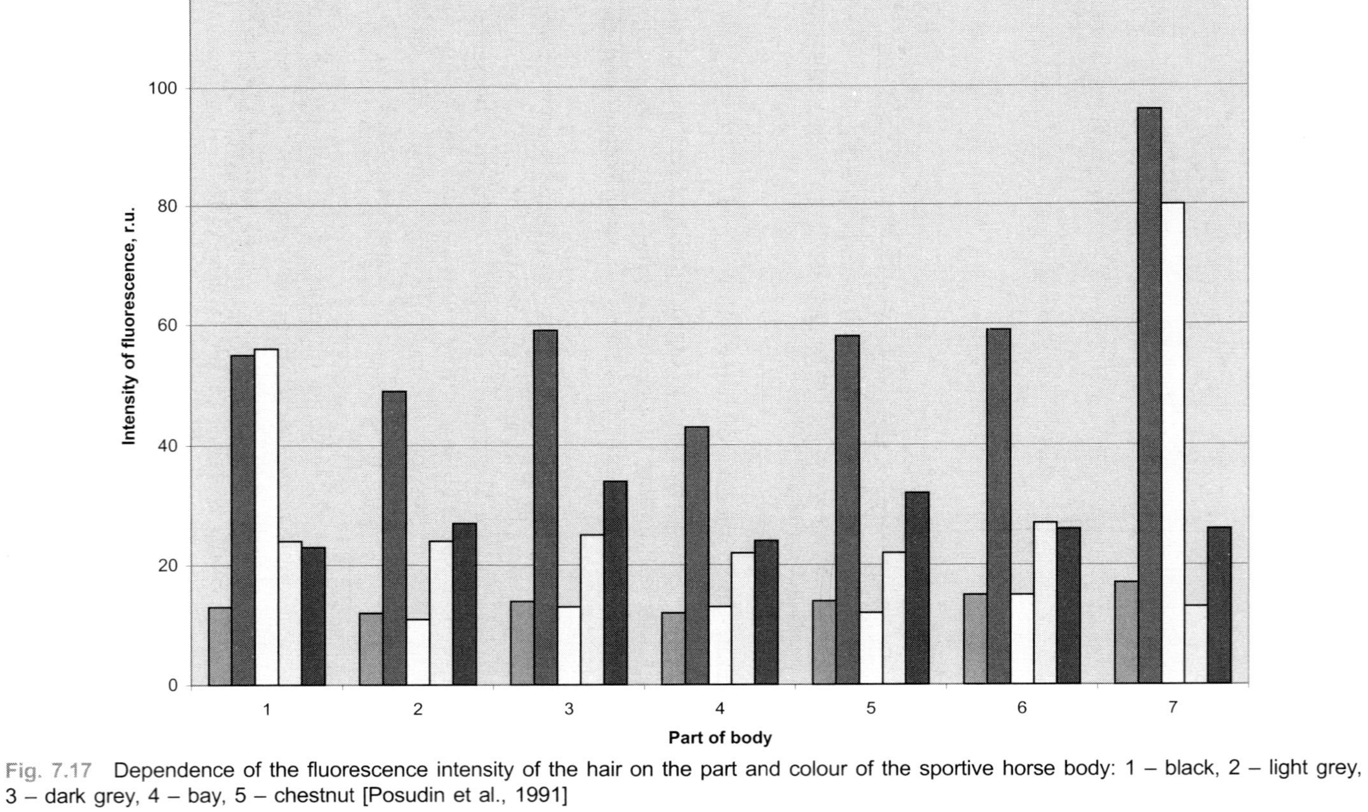

Fig. 7.17 Dependence of the fluorescence intensity of the hair on the part and colour of the sportive horse body: 1 – black, 2 – light grey, 3 – dark grey, 4 – bay, 5 – chestnut [Posudin et al., 1991]

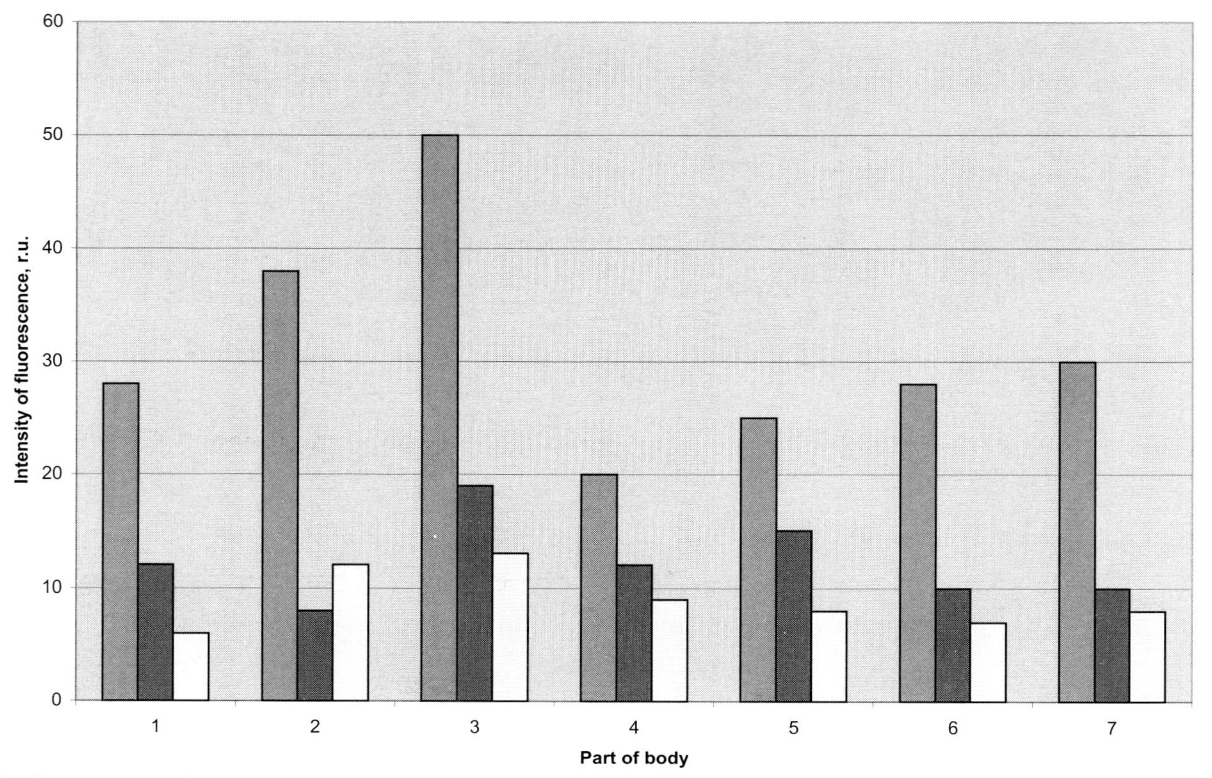

Fig. 7.18 Dependence of the fluorescence intensity of the hair on the part of pony body (stallion of black colouration, mare of chestnut colouration and their foal): 1 – stallion, 2 – mare, 3 – foal [Posudin et al., 1991]

Table 7.1 Dependence of the fluorescence intensity of hair-covering of the Przevalsky horse on parts of the body

Part of Body	*Number of Samples*	*Intensity of Fluorescence*
Cheek	63	44.09 ± 4.85
Neck	68	49.84 ± 4.53
Shoulder blade	69	55.36 ± 4.11
Back	72	61.19 ± 5.13
Side	72	58.03 ± 5.28
Stomach	66	49.18 ± 5.29
Croup	72	57.29 ± 4.88
Extremities	15	23.76 ± 2.38
Mane	54	16.07 ± 1.97
Tail	57	22.11 ± 2.71

These results correlate first of all with colouration of the Przevalsky horse: in spite of diversity of types and varieties, this horse shows definite tendencies in colour of hair-covering (Table 7.2).

Table 7.2 Coefficient of correlation of fluorescence intensity between different parts of the body of Przevalsky horse

Part of Body	*Coefficient of Correlation*	t_d	*P*
Cheek-back	0.590	4.300	>0.999
Cheek-back	0.370	2.210	>0.950
Cheek-croup	0.390	2.560	>0.950
Cheek-back	0.400	2.660	>0.950
Neck-croup	0.530	3.990	>0.999
Neck-back	0.460	3.150	>0.900
Shoulder blade-croup	0.530	5.990	>0.999
Shoulder blade-stomach	0.350	1.990	>0.950
Shoulder blade-mane	0.400	2.300	>0.950
Back-croup	0.680	6.080	>0.999
Back-side	0.430	3.140	>0.900
Back-stomach	0.440	3.020	>0.900
Back-tail	0.350	2.870	>0.950
Side-stomach	0.570	4.440	>0.999
Side-croup	0.410	3.080	>0.900
Side-mane	0.470	3.390	>0.900
Side-tail	0.380	2.700	>0.950
Stomach-extremities	0.610	4.030	>0.999
Stomach-tail	0.780	7.890	>0.999
Stomach-croup	0.410	2.860	>0.900
Stomach-mane	0.440	2.960	>0.900
Croup-mane	0.440	3.060	>0.900
Croup-extremities	0.380	2.310	>0.950
Extremities-tail	0.460	2.920	>0.900

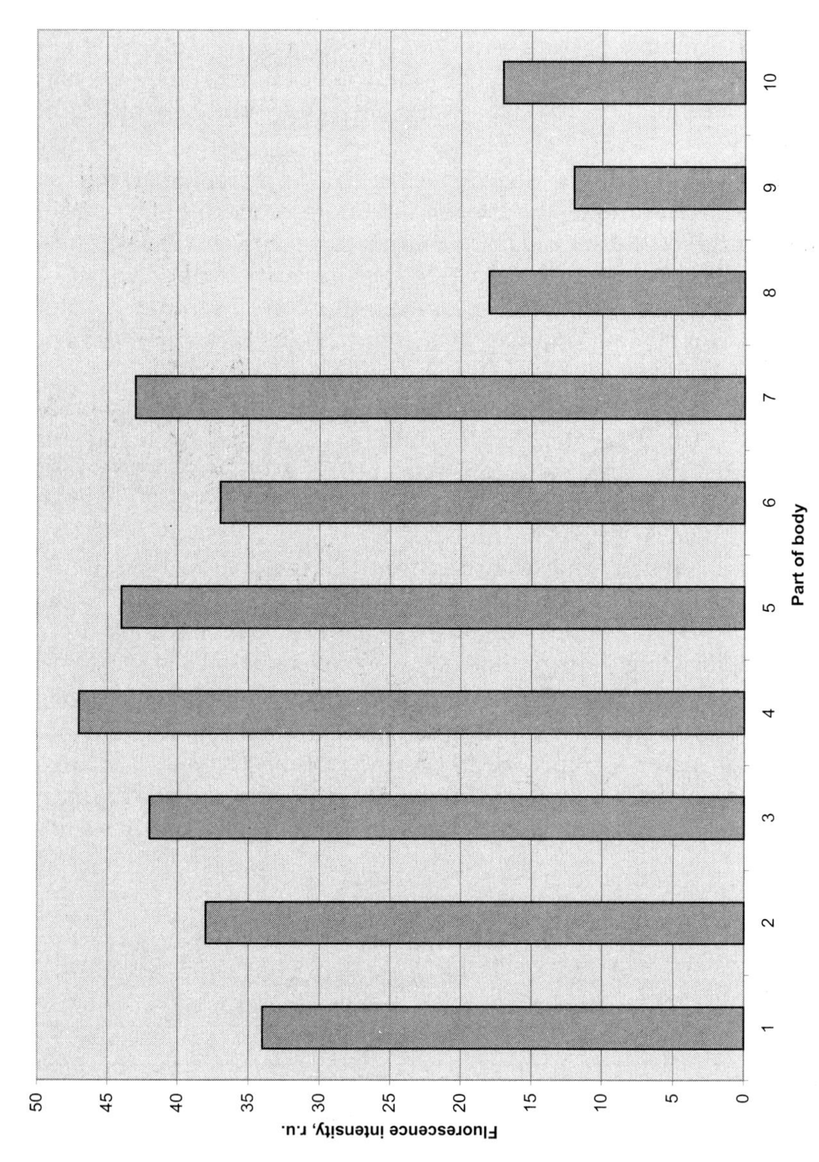

Fig. 7.19 Dependence of the fluorescence intensity of the hair on the part of Przevalsky horse: 1 – cheek, 2 – neck, 3 – shoulder-blade, 4 – back, 5 – side, 6 – stomach, 7 – croup, 8 – extremities, 9 – mane, 10 – tail [Posudin et al., 1998]

The intensity of fluorescence is related to the colouration of hair-covering, for example, stallion "Vidny" demonstrates intensity of fluorescence 17 ± 1 for black hair, 143 ± 2 for white hair, 252 ± 10 for chestnut hair; stallion "Polkan" 31 ± 7 for black hair and 127 ± 9 for white one. The correlation between intensity of fluorescence and colouration of the hair can be explained by different quantity of hair melanine – the hair of dark colouration absorbs light more intensively and re-emits light more weakly.

We did not establish the dependence of hair fluorescence on the sex of the animal, for example, the intensity of fluorescence of the hair from the side was 58.86 ± 6.69 for mare (number of samples 42) and 58.86 ± 8.68 for stallion (number of samples 30); the intensity of fluorescence of the hair from neck was 58.62 ± 6.33 for mare (number of samples 39) and 38.01 ± 35.33 for stallion (number of samples 29) the difference was not statistically significant.

The results of the experiments demonstrated the dependence of fluorescence intensity of the hair-covering on the season – the value of this parameter increased in summer from 5.33 ± 0.33 to 11.67 ± 0.88 (t_d = 6.81; $P > 0.99$) for neck hair, from 9.67 ± 2.19 to 20.67 ± 6.67 (t_d = 7.5; $P > 0.99$) for hair on the back, from 2.00 ± 0.00 to 2.67 ± 0.39 (t_d = 2.03; $P > 0.9$) for mane hair, and decreased from 25.00 ± 0.33 to 16.33 ± 1.53 (t_d = 3.57; $P > 0.95$) for hair on the side and from 22.00 ± 1.25 to 5.67 ± 0.67 (t_d = 11.5; $P > 0.999$) for side hair. The difference between the intensity of fluorescence for other parts of the body was not statistically significant.

It was interesting to study the dependence of fluorescence parameters on the age of the animal. We have studied the age change of fluorescence of the hair of mare "Vorkuta" at the period of time 17 days, 4 months 22 days, and one year 4 months. The hair of the young horse (17 days) demonstrates the maximal values of the intensity of fluorescence (Table 7.3).

Table 7.3 Age changes of fluorescence of hair of Przevalsky horse

Age	*Date of Experiment*	*Intensity of Fluorescence of the Hair from Shoulder Blade, r.u.*			
		n	$I \pm \sigma^2$	t_d	*P*
17 Days	01.06.1993	6	20.00 ± 2.73	0.03	<0.95
4 Months 23 days	14.10.1993	6	11.17 ± 1.01	1.52	>0.90
1 Year 3 months	01.09.1994	6	11.67 ± 0.80	3.49	>0.95

Determination of genotypic indices. The main objective of these experiments was possible application of fluorescence parameters of animal (pony and Przevalsky horse) hair as genotypic signs of animal diversity. The author's and his colleague's attention was drawn to the hair of Scotch pony-stallion "Medok" of black colouration, Estonian pony-mare "Damy" of chestnut colouration and their foal of one year old [Posudin et al., 1993]. The investigation of fluorescence parameters of horse hair as genotypic signs was continued with the hair-covering of the Przevalsky horse. The hair samples were taken from stallion "Vizor 313", mare "Volga 244", and foal "Vetka 490" [Posudin et al., 1998].

The spectral characteristics (intensity of fluorescence) of the hair samples which were taken from different parts of the body of the pony were investigated. The following ratio was used for determination of predomination of genetic levels of the posterity [Plokhinsky, 1964]:

$$F = \frac{S + kM}{1 + k} \quad \text{(Eqn. 7.6)}$$

where k is the coefficient which is determined experimentally as:

$$k = \frac{S - M + 2\Delta}{S - M - 2\Delta} = \frac{M - F}{F - S} \quad \text{(Eqn. 7.7)}$$

$$D = F - \frac{M + S}{2} \quad \text{(Eqn. 7.8)}$$

Here symbols S and M correspond to the fluorescence intensity of the samples which were taken from hair-covering of stallion and mare correspondingly; k is a level of increasing of genetic influence of parents; Δ is exceeding of mean level of the descendant sign over half-sum of mean levels of parents.

The results of measurement of fluorescence intensity of hair which were taken from different parts of the body are presented in Table 7.4 (pony) and Table 7.5 (Przevalsky horse) correspondingly.

It is shown that fluorescence intensity of the foal hair takes the intermediate place between the values of the fluorescence intensity of the parent's hair for pony (Fig. 7.18) and for the Przevalsky horse (Fig. 7.20).

Application of the expression (Eqn. 7.6) made it possible to elucidate which parent – stallion or mare – provides dominating genetic information. The results of calculation are presented in Table 7.6 for pony and in Table 7.7 for the Przevalsky horse.

Table 7.4 Dependence of fluorescence intensity of pony hair on parts of the pony body

Part of Body	$I \pm \sigma^2$		
	Parents		
	Stallion	*Mare*	*Foal*
Head	30.67 ± 2.90	5.67 ± 1.20	8.33 ± 4.70
Neck	22.67 ± 3.10	10.00 ± 0.80	13.00 ± 4.10
Side	27.00 ± 4.30	9.33 ± 1.90	17.00 ± 3.50
Croup	30.33 ± 6.90	7.00 ± 0.80	11.67 ± 0.90
Extremities	32.00 ± 2.40	9.33 ± 1.20	10.67 ± 0.90
Tail	56.67 ± 3.40	15.00 ± 1.00	21.00 ± 1.60

Table 7.5 Dependence of fluorescence intensity of pony hair on parts of the Przevalsky horse body

Part of Body	$I \pm \sigma^2$		
	Parents		
	Stallion	*Mare*	*Foal*
Cheek	48.00 ± 2.80	164.00 ± 1.90	84.00 ± 5.80
Back	80.00 ± 3.90	170.00 ± 0.90	126.00 ± 4.90
Side	66.00 ± 4.20	155.00 ± 2.60	128.00 ± 3.80
Croup	103.00 ± 6.90	62.00 ± 2.80	121.00 ± 5.90
Stomach	84.00 ± 3.50	103.00 ± 1.20	99.00 ± 3.00
Mane	17.00 ± 2.50	20.00 ± 1.00	8.00 ± 0.90
Tail	24.00 ± 2.80	32.00 ± 4.30	25.00 ± 2.70

Table 7.6 Effect of non-additive domination on genotype signs for pony family

Part of Body	*k*	Δ	*Formation of Genetic Level of Descendant*	*Coefficient of Regression for Mare*	*Coefficient of Regression for Stallion*
Head	0.12	−9.84	$F = (M + 0.12S)/1.12$	0.89	0.11
Neck	0.31	−3.33	$F = (M + 0.31S)/1.31$	0.76	0.24
Side	0.77	−1.17	$F = (M + 0.77S)/1.77$	0.56	0.44
Croup	0.25	−7.00	$F = (M + 0.25S)/1.25$	0.80	0.20
Extremities	0.06	−10.00	$F = (M + 0.06S)/1.06$	0.94	0.06
Tail	0.17	−14.84	$F = (M + 0.17S)/1.17$	0.85	0.15

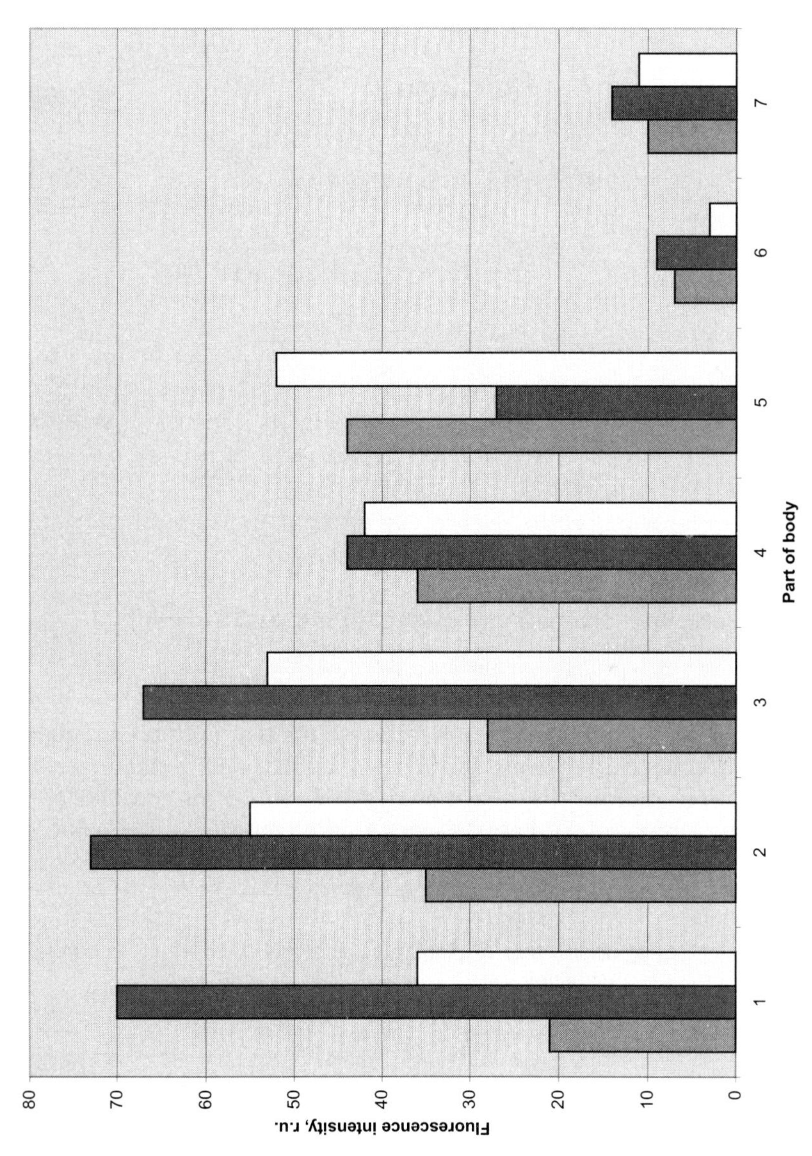

Fig. 7.20 Dependence of the fluorescence intensity of the hair on the part of Przevalsky horse body (1 – stallion, 2 – mare, 3 – foal) [Posudin et al., 1998]

Table 7.7 Effect of non-additive domination on genotype signs for the Przevalsky horse family

Part of Body	*k*	*Δ*	*Formation of Genetic Level of Descendant*	*Coefficient of Regression for Mare*	*Coefficient of Regression for Stallion*
Cheek	2.22	–22.00	$F = (M + 2.22S)/3.22$	0.31	0.69
Back	0.96	1.00	$F = (M + 0.96S)/1.96$	0.51	0.49
Side	0.44	17.50	$F = (M + 0.44S)/1.44$	0.69	0.31
Croup	–3.30	38.50	$F = (M – 3.3S)/ – 2.30$	–0.43	1.43
Stomach	0.27	5.50	$F = (M + 0.27S)/1.27$	0.790	0.21
Mane	–1.34	–10.50	$F = (M – 1.34S)/ – 0.34$	–2.90	3.90
Tail	7.00	–30	$F = (M + 7S)/1.70$	0.59	4.12

Method of microspectrofluorometry of horse hair can be used for quantitative evaluation of level and character of pigment distribution within hair, studying internal structure of hair, diagnostics of state of horse under influence of feeding conditions and effects of external factors, and degree of heredity [Posudin, 1988].

It is very informative to retrace the level of the heredity of hair fluorescence by several generations of animals.

7.4.3 Laser Diffractometry of Animal Hair and Nasolabial Mirror

Hair of Sportive Horses. Laser diffractometry of horse hair, which was taken from different parts of the animal body, made it possible to relate parameter l_m (interval between diffraction maxima) and d (diameter of hair). The application of dispersive analysis of two-factor complex (i = part of body, j = colour of hair) testifies to the fact that the calculated criteria F_i, F_j and F_{ij} are higher than the critical value F_c for the level of significance P = 0.05 and P = 0.01 (Table 7.8).

Table 7.8 Results of application of dispersive analysis of two-factor complex (*i* – part of body, *j* – colour of hair)

Source of Variation	*Degree of Freedom*	*Actual Factor F_a*	*Critical Factor F_c (P = 0.05)*	*Critical Factor F_c (P = 0.01)*
Factor *i* (part of body)	7	$F_i = 25.987$	2.200	3.000
Factor *j* (colour of hair)	2	$F_j = 19.487$	3.200	5.000
Common action of two factors *ij*	14	$F_{ij} = 13.108$	1.800	2.200
Residue	60	$F_z = 1$	–	–

Independent and common effects of both factors and their interaction are statistically significant as soon as $F > F_c$.

Vibrissa of Calves and Heifers. The results of laser diffractometry of the vibrissa of bull calves and heifers were compared with the weight of the calves: 371 heads were distributed to nine weight ranks (from 26–27 kg to 42–43 kg). Analysis of diffraction patterns (Figs. 7.21–7.28) demonstrated that the dominating weight groups were [Posudin and Trofimenko, 1994]: 28–29 and 30–31 kg in heifers; and 30–31 and 32–33 kg in bull calves. The corresponding thickness of vibrissa consists of: 100–150 μm (Δl = 40–60 mm) for 28–29 kg group of heifers; 85–120 μm (Δl = 50–70 mm) for 30–31 kg group of heifers and bull calves; and 80–122 μm (Δl = 49–75 mm) for 32–33 kg group of bull calves.

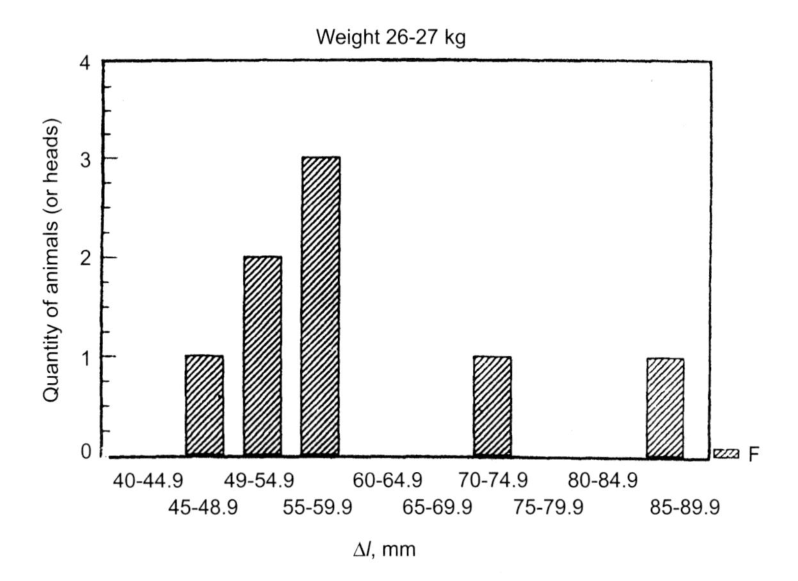

Fig. 7.21 Results of laser diffractometry of the vibrissa of heifers of weight 26-27 kg. Here: Δl – distance between diffraction maxima [Posudin et al., 1993]

Thus, the method of laser diffractometry can possibly help to estimate the relationship between the hair thickness, weight and sex of the animals.

Another application of laser diffractometry is related to the analysis of the hair orientation. It is known that the angle of hair inclination in animals depends on the environmental temperature: this angle increases from 28° to 63°C with the decrease in temperature from +11° to –24°C;

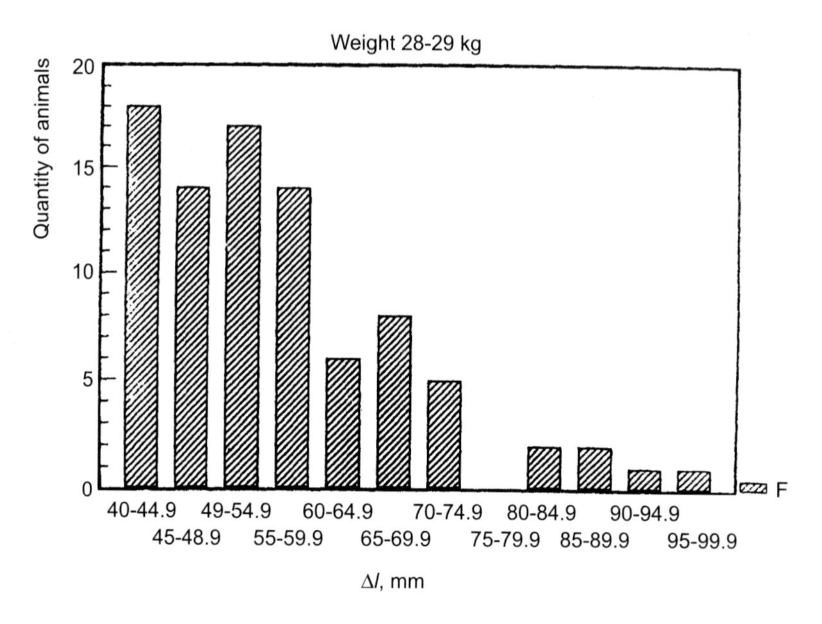

Fig. 7.22 Results of laser diffractometry of the vibrissa of heifers of weight 28-29 kg [Posudin et al., 1993]

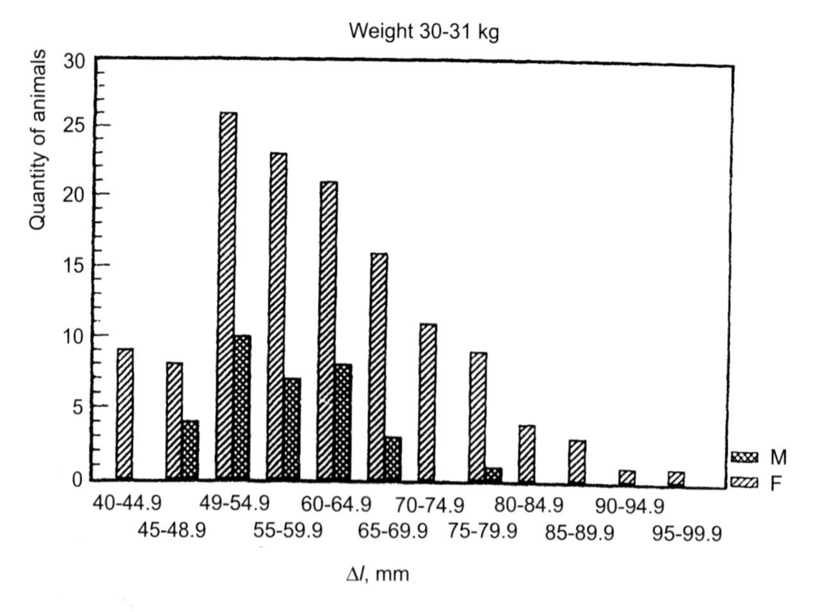

Fig. 7.23 Results of laser diffractometry of the vibrissa of bull calves and heifers of weight 30-31 kg. Here and in Figs. 7.24, 7.25, and 7.28: *M* – calves, *F* – heifers, Δl – distance between diffraction maxima [Posudin et al., 1993]

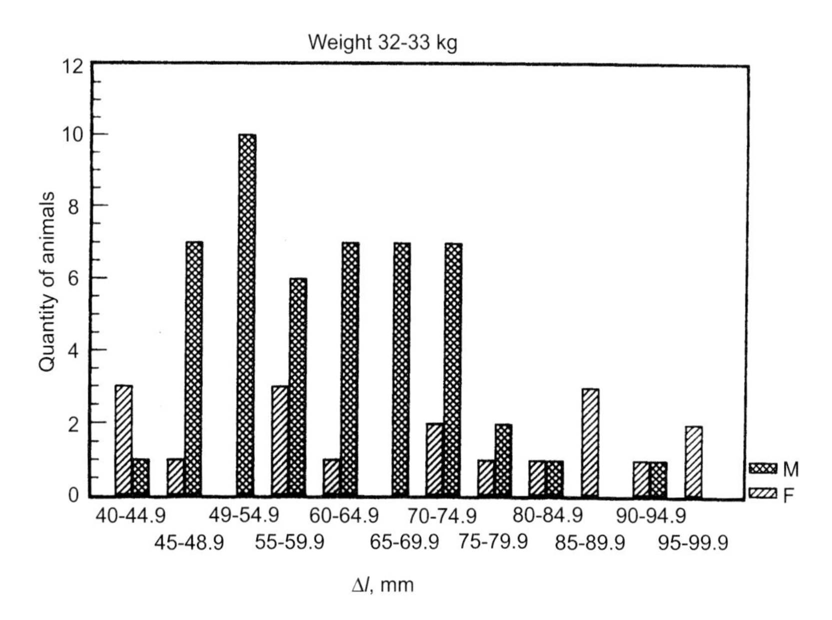

Fig. 7.24 Results of laser diffractometry of the vibrissa of bull calves and heifers of weight 32-33 kg [Posudin et al., 1993]

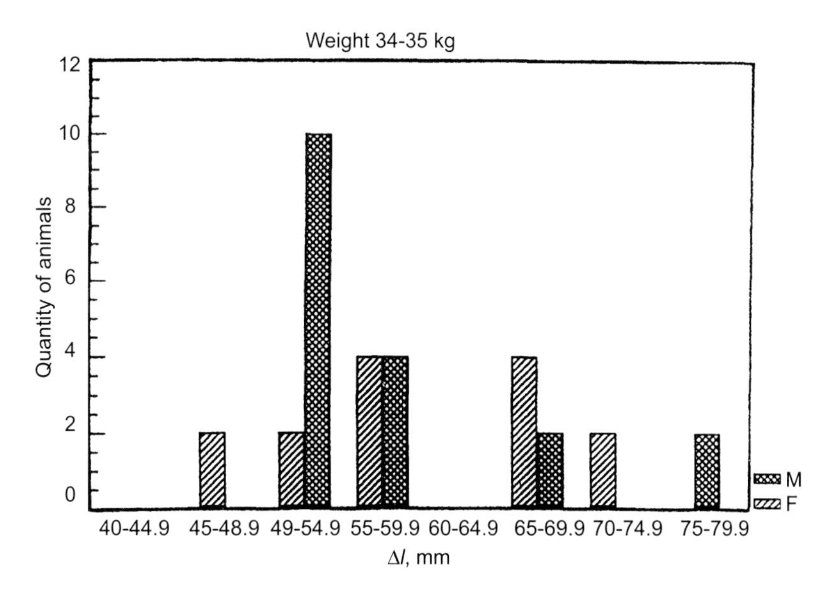

Fig. 7.25 Results of laser diffractometry of the vibrissa of bull calves and heifers of weight 34-35 kg. *M* – calves, *F* – heifers, Δl – distance between diffractional maxima [Posudin et al., 1993]

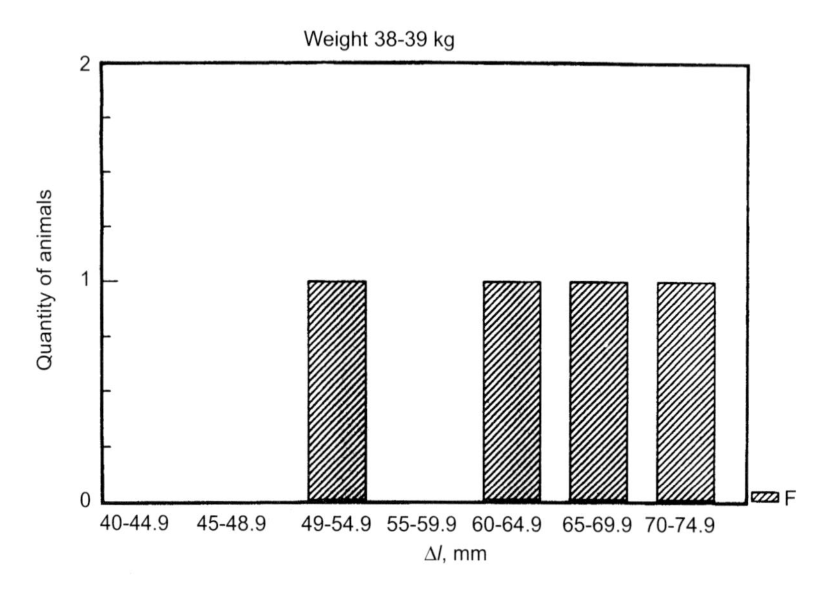

Fig. 7.26 Results of laser diffractometry of the vibrissa of heifers of weight 38-39 kg [Posudin et al., 1993]

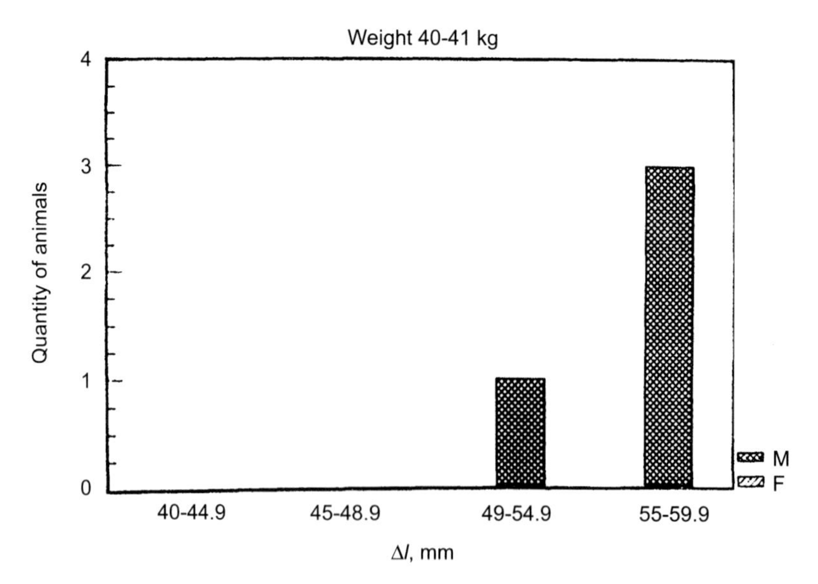

Fig. 7.27 Results of laser diffractometry of the vibrissa of bull calves of weight 40-41 kg [Posudin et al., 1993]

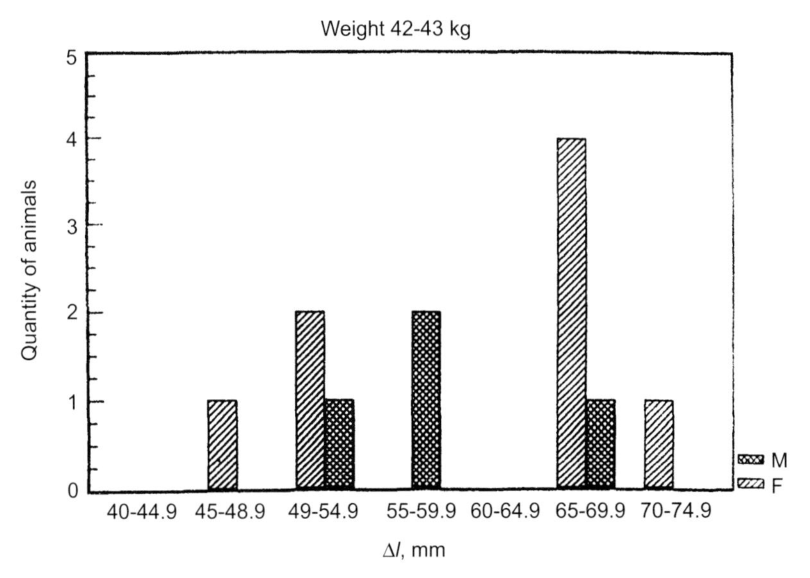

Fig. 7.28 Results of laser diffractometry of the vibrissa of bull calves and heifers of weight 42-43 kg [Posudin et al., 1993]

the thickness of the hair layer changes correspondingly from 11 to 21 mm. However, the change in the angle of inclination provokes the alteration of spatial orientation of the diffraction pattern (Fig. 7.29). The measurement of the angular position of the hair and diffraction pattern helps to evaluate the effects of thermal stresses on farm animals.

7.5 SELECTION IN CATTLE-BREEDING

Nasolabial Mirror of the Cattle. The structure of nasolabial mirror of the cattle is characterized by a specific pattern which presents a peculiar fingerprint of the animal. There are several different types of dermatoglyphs such as *crown, twig, ear, grain* and their combinations [Trofimenko et al., 1990]. The analysis of the structure of this dermatoglyph is promising from the point of view of early prognosis of the animal.

The differentiation of the cattle is performed by the analysis of genealogical properties of external, morphological, biochemical, and physiological indices.

The skin-coat of the animal presents the important factor which could be used for early prognosis of the main trends of cattle development. The thickness of the skin, its density and friability depend on the sex, age, constitution of the animal, conditions of feeding and maintenance.

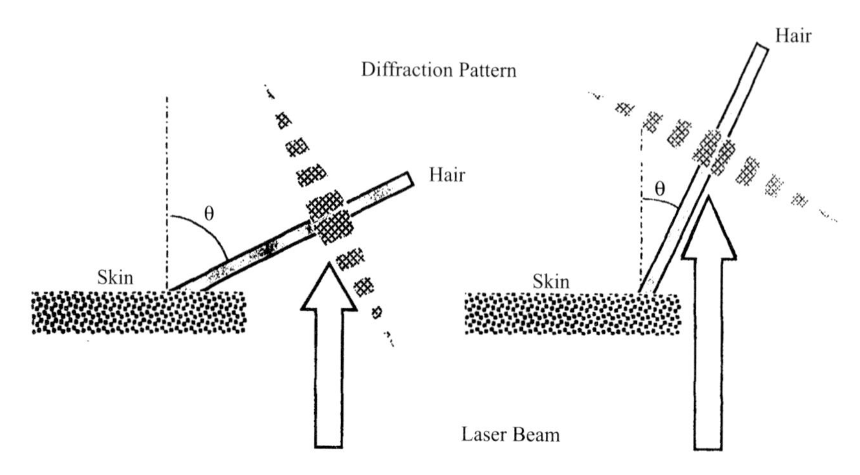

Fig. 7.29 Dependence of spatial orientation of diffraction pattern on the inclination angle θ of the hair [Posudin, 1998]

Considering the automation of laser systems of dermatoglific identification and classification, the first variant [Trofimenko et al., 1990] consists of laser, Fourier-objective, optical system (diaphragm and mirrors), videocamera and system of elaboration of data. The radiation of laser passes through the slide with the picture of nasolabial structure (Fig. 7.30). The analysis of spatial structure of the optical image can give the information about of the sex of the individual.

The second approach is based on laser diffractometry of nasolabial mirror [Posudin et al., 1998]. The procedure of measurements includes the interaction of laser radiation with the quasi-periodical structure of the mirror, analysis of diffraction pattern (Fig. 7.31) and identification of animal diversity.

7.6 LASER CONTROL OF BIRD FEATHERING

Morphology of bird feathering, its anatomical, histological properties and colour depend on the state, sex and age of the bird, the external factors, season of year, conditions of maintenance and feeding of the birds.

The structure of the bird feather is presented in Fig. 7.32*a*. The main parts of the feather are pivot (*scapus*) and large fan (*vexillum*). Scapus consists of *calamus* and *rachis*. *Vexillum* includes *rami* (singular: *ramus*), the central shaft of feather barbule that is attached to the rachis, and

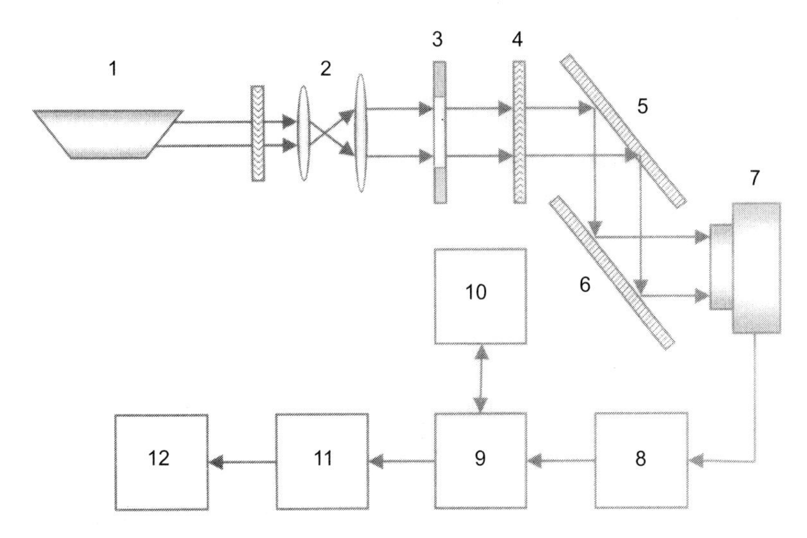

Fig. 7.30 Automated laser system of dermatoglific identification and classification. 1 – laser; 2 – Fourier-objective; 3 – diaphragm; 4 – film; 5, 6 – mirrors; 7 – videocamera; 8 – input controller; 9 – computer; 10 – control panel; 11 – block of conjugation; 12 – display [Trofimenko et al.,1990]

dozens of small *radi* (singular: *radius*) that extend out from either side of the rami.

The estimation of morphology of the feather is performed as per the following parameters [Svetozarov and Shtraikh, 1939]: the length A of the feather part of vexillum; the length B of the lower part of vexillum; the total length $A+B$ of vexillum; the length C of scapus; the width D of vexillum; the diameters d_1 , d_2 , d_3 of scapus in different places; the points E, F, G, H and O, which are situated 5 mm from the scapus and where the number of rami is determined.

These parameters can be used as the taxonomical indices which are related to the species and race of the bird (Table 7.9).

As a whole, the feather presents the quasi-periodical structure which can be considered a peculiar diffraction grating. Laser radiation can interact with such a structure and produce a diffraction pattern (Fig. 7.32*b*). The relative intensity and spatial position depends on the geometry of the feather (diameter of the *ramus* and *radius*) which reflects the healthy status of the bird, diseases, regime of feeding, etc. [Posudin and Lepeshenkov, 1991].

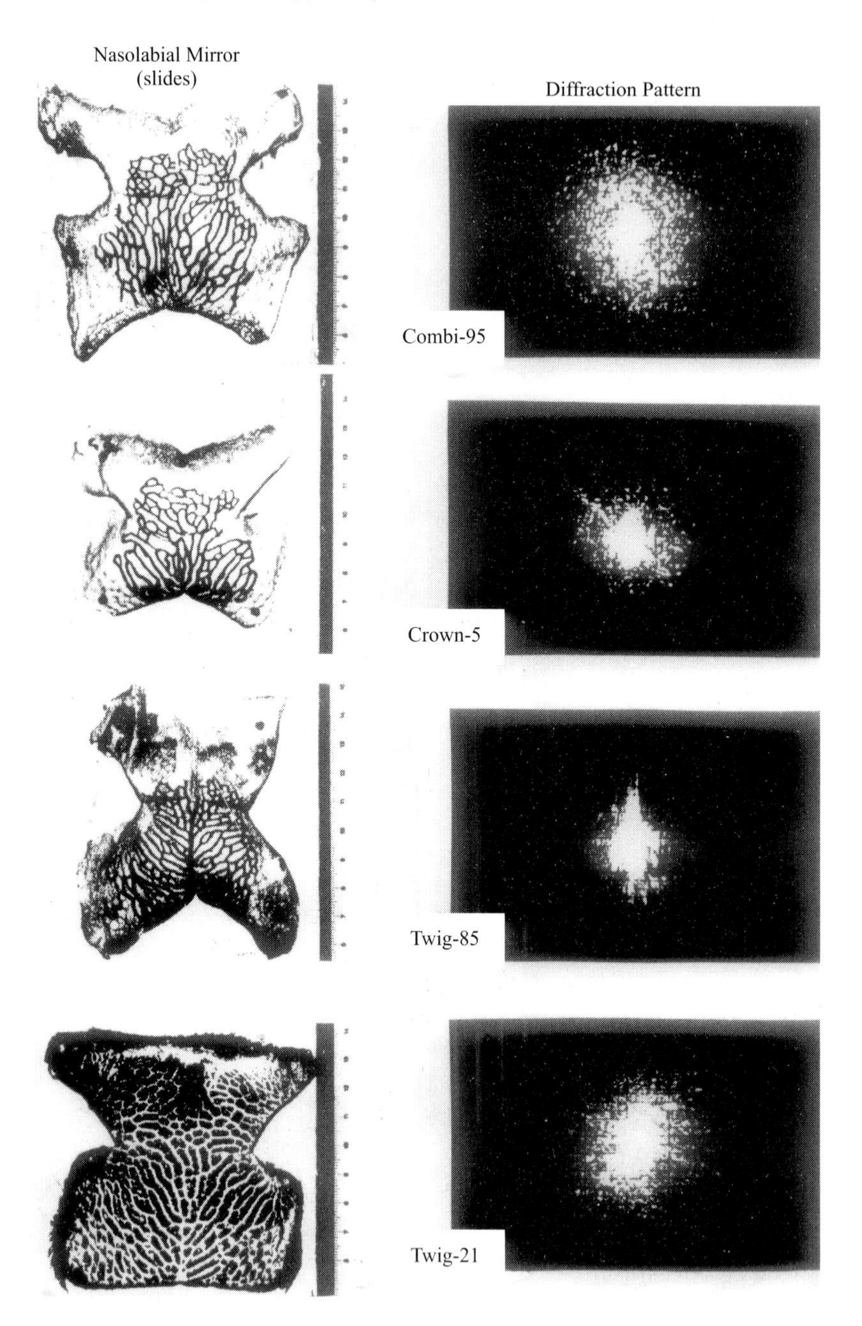

Fig. 7.31 The interaction of laser radiation with the quasi-periodical structure of the nasolabial mirror of the cattle, analysis of diffraction pattern and identification of animal diversity [Posudin et al., 1998]

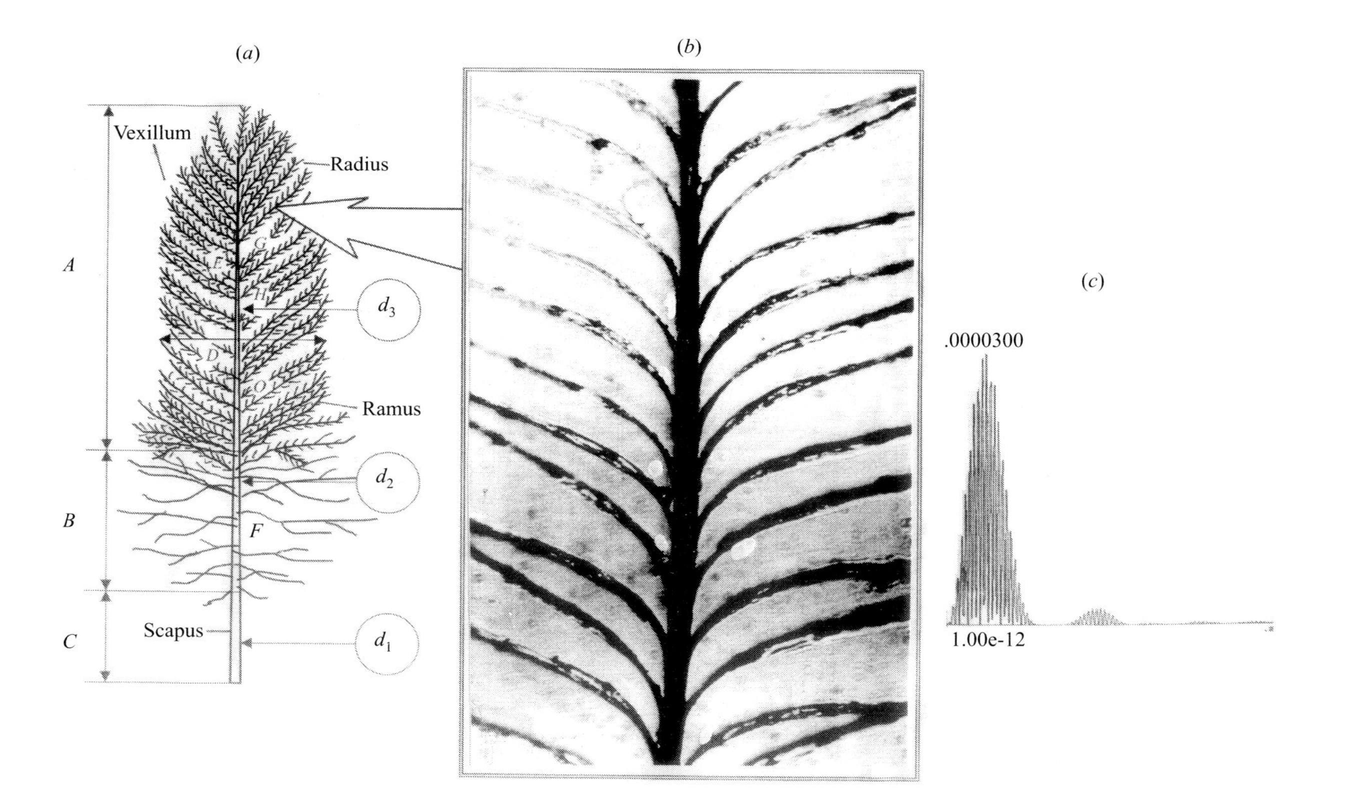

Fig. 7.32 The bird feather [Svetozarov and Shtraikh, 1939]: *a* – main parameters of feather (explanation in text); *b* – picture of the ramus and radius; *c* – calculated diffraction pattern which is caused by the interaction of laser radiation with structural elements (ramus and radius) of the feather [Posudin and Lepeshenkov, 1991]

Table 7.9 Specific and racial distribution in the size of the main feather parts [Svetozarov, Shtraikh, 1939]

Indices	*Goose*	*Ducks*	*Ducks*	*Hen*
		Pekin	*Rouen*	
Whole length of feather, mm	60	55	46	70
Length, mm				
• pivot	10	8	10	6
• fan	90	92	90	94
• feather part of fan	64	68	68	44
• down part of fan	28	25	22	59
Width of fan, mm	52	45	42	38

7.7 SUMMARY

The dependence of transmittance of horse hair on the illuminance E was investigated in the author's experiments. It was shown that the curves $\tau = f(E)$ depend strongly on the colour and part of the animal body; each curve is characterized by a linear part of dependence $\tau = f(E)$, and the region of saturation where $\tau \rightarrow 1$. The hair of grey colouration has maximal transmission, while minimal transmission is typical for the hair of black colouration. The linear dependence $\tau = f(E)$ can be used for quantitative estimation of colour of hair-covering of horses.

The fluorescence intensity of the hair depends on the part and colour of the body (sportive horses, ponies and Przevalsky horse). Method of microspectrofluorometry of the horse hair can be used for quantitative evaluation of level and character of pigment distribution within the hair, studying internal structure of the hair, diagnostics of state of the horse under influence of feeding conditions, and effects of external factors.

It is very attractive to use fluorescence parameters of the animal (pony and Przevalsky horse) hair as genotypic signs of animal diversity for determining the degree of heritability.

7.8 REFERENCES

Bannikov, A.C. 1958. Distribution geographique et biologie du cheval sauvage et chameau de Mongolie. *Mammalia,* 22: 152–160.

Born, M. and Wolf, E. 1970. *Principles of Optics.* 4th ed. Pergamon Press, Oxford, London, N.Y.

Combs, D.K. 1987. Hair analysis as an indicator of mineral states of livestock. *J. Amer. Sci.* 65(6): 1753–1758.

Durst, W. 1963. *Exterior of Horse.* Moscow-Leningrad. Selkhozgiz. p 552.

Gopka, B.M., Kalantar, O.A. and Pavlenko, P.M. 1989. *Horses in Àgriculture*. Kiev, Urozhai, p 150.

Groves, C.P. 1986. The taxonomy, distribution and adaptations of recent Equids. In: *Equids in the Ancient World*. R.H. Meadow and H.P. Uerpmann, eds. Dr Ludwig Reichert Verlag, Wiesbaden, Germany, p.11-65.

Ioganson, I., Rendel, Y. and Gravert, O. 1970. *Genetics and Breeding of Domestic Animals*. Moscow, Kolos, Russia (or USSR) p 351.

Klimov, V.V. 1990. *The Przevalsky Horse*. Agropromizdat, Moscow, p 253.

Kostaneski, W. 1970. *Choroby wlosów*. Warsaw, Poland.

Plokhinsky, N.A. 1964. *Heritability*. Novosibirsk, Siberian Branch of Academy of Science, p 194.

Poliakov, I.S. 1881. Przewalski's Horse (*Equus przwalskii* n. sp.). *Izvestia Russki Geographicheski obschestva*, St. Petersburg, 17: 1–20.

Posudin, Yu.I. 1995. *Biophysics*. Urozhaj, Kiev. p 222.

Posudin, Yu.I. 1988. *Lasers in Agriculture*, Science Publishers, Inc. Enfield, NH, USA. p 188.

Posudin, Yu.I., Chervinsky, L.S., Yasynetska, N.I. and Zharkykh, T.L. 1998. Investigation of fluorescence characteristics of Przevalsky horse hair. *Sci. Herald of Nat. Agr. Univ.* 4: 3–9.

Posudin, Yu.I., Gopka, B.M. and Zatvarsky, B.M. 1991. Optical, spectral and laser methods of investigation of horse hair cover. *Biol. Sciences*. Depon VINITI, N257–B91, p 28.

Posudin, Yu.I. and Lepeshenkov, V.F. 1991. Optical and laser methods of poultry-breeding control. *St.-Petersbourg Agr. Univ. Publ.* 1: 59–64.

Posudin, Yu.I., Trofimenko, A.L., Koval, Y.M. and Palecha, N.P. 1993. Spectral characteristics of animals' hair as genetic indices. *Cytology and Genetics*. 27: 74–77.

Posudin, Yu.I. and Trofimenko, A.L. 1994. Laser diffractometry of calves hair. *Bull. Agr. Science* 9: 85–89.

Posudin, Yu.I., Trofimenko, A.L. and Vinnichuk, D.T. 1998. Method of identification of animal features through the structure of nasolabial mirror. Patent of Ukraine 25105A; N 95041577, 1998.

Seiji, M. 1973. On the melanization process of melanosomes. *Pigment Cell*. 1: 39–46.

Sokolov, E.V. and Zhenevskaya, R.P. eds. 1988. *Guidance to study skin-covering of mammals*. Moscow, Nauka, p 280.

Svetozarov, E.A. and Shtraikh, G.G. 1939. Comparative morphology of feather in agricultural birds. *Izv. AN USSR. Ser. Biol* (Russian). 5: 800–822.

Trofimenko, A.L. and Posudin, Y.I. 1998. Description of polyphenism of features during selection of cattle. *Sci. Herald of Nat. Agr. Univ.* 10: 55–61.

Trofimenko, A.L., Vinnichuk, D.T. and Yakovlev, S.Ġ. 1990. Forecast and selection of the cattle. In: *Agriculture*. Series Cattle-Breeding and Veterinary. Kiev, p 36.

Wojcikowska-Soroczynska, M., Radosz A. and Fiedorczuk E. 1987. Quantitative appearance of pigments in hairs of sheep and of selected animal species. *Ann. Warsaw Agr. Univ.-SGGW-AR Animal Science*, 21: 17-21.

CHAPTER

8

Spectroscopic Analysis of Agronomic Plants

8.1 LASER SPECTROFLUOROMETRY OF AGRONOMIC PLANTS

8.1.1 Photosynthesis and Fluorescence of Chlorophyll

Photosynthesis is the conversion of light energy into stabilized chemical energy through light absorption by a pigment molecule, excitation energy transfer and the photochemical reaction in photosystem *PSII*. The process of deexcitation of the absorbed light energy is related to heat emission and chlorophyll fluorescence. Fluorescence is the radiative process which is based on the transition between electronic states of the same multiplicity; usually it occurs from the ground vibrational state of the first electronic singlet state S_1 to various vibrational levels in ground singlet state S_0, and is accompanied by the emission of photons. The light absorbed by the accessory pigments (chlorophyll *b* and carotenoids) is transferred to chlorophyll *a*. That is why the primary processes of photosynthesis are reflected by chlorophyll *a* fluorescence. The yield of chlorophyll fluorescence depends on numerous factors in a very complex manner; the inverse relationship between *in vivo* chlorophyll fluorescence and photosynthetic activity of leaves can be practically used to study leaf development and detection of stress effects in green plants. Either leaf anatomy and morphology [Sestak, 1985; Sestak and Siffel, 1988; Lichtenthaler and Rinderle, 1988], or various stress conditions [Subhash and Mohanan, 1994] can govern the rate of photosynthesis and provoke the changes of fluorescence parameters. Fluorescence emission can also be studied as a means of assessing chlorophyll content, which in some plant products is related to maturity or certain quality attributes [Kays and Paull, 2004].

8.1.2 Fluorescence Properties of Agronomic Plants

The fluorescence emission spectrum of a green leaf is characterized by a maxima at 735–740 nm and 685–690 nm (red part of spectrum) and at 440–450 nm (blue part); the leaves of some plants demonstrate a shoulder near 520–530 nm (green part). According to recent theory about the character of the fluorescence spectrum of a green leaf, the red maxima originate from chlorophyll *a*. The experimental investigations showed the presence of a number of native forms of chlorophyll. There are the fluorescence bands at room temperature near 672, 677, <u>682, 687, 693–695</u>, 700, 720–725, <u>735</u>, 750–760, <u>800–820</u> (the underlined figures correspond to the intense bands) (Kochubey, 1986). The appearance of these bands is related to the process of aggregation of pigment molecule and formation the chlorophyll–protein complexes.

It is not defined which compounds are responsible for the blue and green parts of the fluorescence spectrum. There is an opinion [Lang et al., 1991; Lichtenthaler et al., 1991; Goulas et al., 1990; 1991] that various phenolic plant constituents in the vacuoles and cell walls of both the epidermal layer and the mesophyll cells such as the coumarines aesculetin and scopoletin, with a very high fluorescence yield, and cinnamic acids and derivatives, with a lower yield (such as caffeic and chlorogenic acid, sinapic acid and catechin) are responsible for the blue-green fluorescence. Some investigators [Chappelle et al., 1991] consider that β-carotene, nicotinamide adenine diphosphate (NADPH), and vitamin K_1 participate in fluorescence at 440 nm. However, purified β-carotene does not show any blue or green fluorescence [Lang et al., 1991]; the yield of blue fluorescence of NAPDPH is extremely low [Lang et al., 1992]; the phylloquinone, K_1, fluoresces only under decomposition after intense UV exposure [Interschick-Niebler and Lichtenthaler, 1981].

The red chlorophyll fluorescence with two maxima near 690 nm and 740 nm emanates from the chlorophyll *a* in the chloroplasts of the leaves' mesophyll cells [Stober and Lichtenthaler, 1993].

The shape and intensity of the main bands in fluorescence emission spectrum of green leaves depend on the pigment content which may vary due to such factors as the growth phase of the plants, natural stresses (high light, heat, water shortage and mineral deficiency) and anthropogenic stresses (agrochemical treatment, air pollution, ozone, acid rains, heavy metals and ultraviolet irradiation).

8.1.3 Plant Samples

The following were investigated during fluorescence measurements:

- Leaves of *cereals*: rye, barley (winter and spring sorts), wheat (winter and spring sorts, soft and stiff species), hybrids amphidiploid, corn.
- Leaves of *vegetables*: cucumber, tomato, haricot, vegetable marrow, pumpkin, pepper, sunflower;
- *Vegetables:* vegetable marrow, tomato, haricot, horse-radish, sorrel.
- *Berries:* currants and raspberry-cane. The sprouts of the cultures were grown in Petri dishes at room temperature in the period of three to eight days. The length of the sprouts varied from 5 to 25 mm. The list of etiolated sprouts (winter and spring sorts) of cereals is given in Table 8.1, etiolated sprouts (winter soft sorts) of cereals in Table 8.2, and of yellow stems (winter and spring sorts) of cereals in Table 8.3.

8.2 INSTRUMENTATION

The schematic representation of laser spectrofluorometer (which was worked out by the author of this work) is given in Fig. 8.1. Nitrogen laser *1* was used as the source of excitation of chlorophyll fluorescence. The main characteristics of this laser are: wavelength of generation 337 nm, frequency of pulse repetition 50 Hz, an average power output of 3 mW, duration of pulse about 10 ns, and divergence of beam $3 \cdot 10^{-3}$ radians.

The laser radiation is directed through the semitransparent glass plate *2* to the sample *3* (a single intact leaf) and to the fluorescent standard *4* (a solution of fluorescent). Either the sample, or the cell with fluorescent standard is oriented at an angle of 45° to the laser beam in order to escape non-desirable absorption of laser radiation by the sample (or standard) volume.

The fluorescence emission of the leaf (or the standard), which is collected by a spherical mirror *5*, passes through the cut-off filter *6*, and is focused on the entrance slit *7* of monochromator *8*. The dispersive element (prism) *9* of this monochromator is linked with the electrical motor *10*, which provides the rotation of the prism and selection of the wavelength. The intensity of fluorescence is detected by photomultiplier *11*; the electrical signal of it is amplified and recorded by the readout system (amplifier *12* and recorder *13*).

Table 8.1 Etiolated sprouts (winter and spring sorts) of cereals

Cereal Species	*Fluorescence Indices*									
	K_1 *(430/460)*	K_2 *(460/530)*	K_3 *(690/735)*	K_4 *(460/690)*	K_5 *(430/690)*	K_6 *(530/690)*	K_7 *(430/735)*	K_8 *(460/735)*	K_9 *(530/735)*	K_{10} *(430/530)*
Rye "Ilmen"	1.0	1.5	3.2	5.8	5.9	3.9	-	-	-	1.5
Rye "Zhitomyrskaya-tetra"	1.0	1.3	3.5	1.3	1.4	0.8	4.7	4.6	2.9	1.3
Rye "Belarusskaya-23"	1.0	1.6	3.7	1.8	1.8	1.1	6.8	6.5	4.1	1.7
Rye "Belta"	1.0	1.4	3.6	1.3	1.3	1.0	4.4	4.6	3.4	1.3
Barley winter "Mirazh"	1.1	1.4	3.1	1.3	1.4	0.9	4.2	3.9	2.8	1.5
Barley winter "Barvynok"	1.1	1.8	4.2	2.9	3.1	1.6	13.2	12.3	6.9	1.9
Barley spring "Crym"	1.1	1.5	4.2	3.0	3.2	2.0	13.6	12.6	8.9	1.6
Barley spring "NAU"	1.1	1.6	3.4	2.0	2.1	4.2	7.2	6.8	4.2	1.7
Wheat soft "Belozerkovskaya-47"	1.0	1.4	3.9	1.4	1.5	1.0	5.7	5.4	3.8	1.5
Wheat winter soft "Ivanovskaya-60"	1.1	1.5	4.0	2.5	2.6	1.6	10.0	9.8	6.4	1.6
Wheat winter soft "Polesskaya-70"	1.0	1.6	4.4	2.7	2.7	1.7	11.9	11.9	7.7	1.6
Wheat winter soft "Kiyanka"	1.0	1.5	3.4	3.2	3.2	2.1	11.0	10.7	7.2	1.5
Wheat winter soft "Kiyanka" (grain)	1.0	1.5	1.5	8.3	8.4	5.6	12.5	12.3	8.3	1.5
Wheat winter soft "Mironovskaya-808"	1.0	1.6	4.6	2.7	2.7	1.8	12.3	12.5	8.1	1.5
Wheat winter stiff "Iceberg odessky"	1.0	1.6	4.1	1.6	1.5	1.0	6.2	6.4	4.1	1.5

(Table Contd.)

(*Table Contd.*)

Wheat winter stiff “Korund”	0.9	1.6	3.6	2.1	1.9	1.3	6.9	7.4	4.5	1.5
Wheat winter stiff “Parus”	0.9	1.5	3.7	1.1	1.0	0.8	3.8	4.1	2.8	1.3
Wheat winter stiff “Svetlana”	1.0	1.5	3.5	2.5	2.4	1.9	8.2	7.9	5.3	1.5
Wheat spring stiff “Kharkov-7”	1.0	1.5	3.1	2.4	2.3	1.6	7.3	7.6	5.0	1.5
Wheat spring soft “Diana”	1.0	1.6	3.6	2.6	2.5	2.0	8.4	8.0	5.2	1.5
Wheat spring soft “Vector”	1.1	1.5	4.0	1.6	1.7	1.1	6.8	6.5	4.4	1.6
Hybrid amphidiploid “Diploid-2067”	1.0	1.6	3.4	1.6	1.6	1.0	5.4	0.6	3.5	1.6
Hybrid amphidiploid “Mal-1”	1.0	1.5	2.9	2.6	2.5	1.7	7.1	7.4	4.9	1.5
Hybrid amphidiploid “Diploid 3/5”	1.0	1.5	4.1	2.7	2.6	1.9	10.7	11.1	7.7	1.4
Corn line R-154	1.0	1.7	2.2	20.5	20.9	2.14	5.24	4.3	24.2	1.7
Corn “Collectivny 101”	0.9	1.6	4.4	6.2	5.6	4.0	24.6	27.3	17.5	1.4
Corn hybrid “Moldavsky”	1.0	1.4	1.4	15.8	16.3	11.1	22.2	21.6	15.2	1.5

Table 8.2 Fluorescence indices of different species of etiolated sprouts of winter soft wheat

Species	K_2 (460/530)	K_3 (690/735)	K_4 (460/690)
"Belozerkovskaya-47"	1.4	4.4	1.0
	1.7	4.1	1.3
	1.4	3.4	0.8
	1.5	4.2	1.6
	1.5	4.3	1.6
	1.4	3.9	1.4
"Polesskaya-70"	1.5	4.3	2.0
	1.7	3.3	5.9
	1.6	4.1	3.1
	1.6	4.4	4.9
	1.6	3.8	4.7
	1.6	4.4	2.7
"Kiyanka"	1.5	4.0	3.6
	1.6	4.4	1.9
	1.5	3.4	3.2
"Ivanovskaya-60"	1.6	4.6	2.9
	1.8	4.6	2.6
	1.9	4.4	2.8
	1.5	4.0	2.5
"Mironovskaya-808"	1.6	4.3	3.1
	1.6	4.6	2.7

8.3 RESULTS OF LASER SPECTROFLUOROMETRY OF A SINGLE LEAF

8.3.1 Fluorescence Indices

The following fluorescence intensity ratios have been used as the vegetation indices [Ivanitskaya and Posudin, 1989]: K_1 = 430/460, K_2 = 460/530, K_3 = 690/735, K_4 = 460/690, K_5 = 430/690, K_6 = 530/690, K_7 = 430/735, K_8 = 460/735, K_9 = 530/735, K_{10} = 430/530.

8.3.2 Fluorescence Emission Spectra during Greening of Etiolated Leaves

The values of the fluorescence indices for the etiolated sprouts of cereals (soft and stiff species, winter and spring sorts) are presented in Tables 8.1 and 8.2. It is shown that the values of $K_1 \approx 1$, $K_2 - K_{10} \approx 1.5$, and $K_3 \approx 3$–4 are constant. However, the ratio of blue/red bands is different.

Table 8.3 Fluorescence indices of yellow stems of cereals

Cereal Species	*Fluorescence Indices*									
	K_1 (430/460)	K_2 (460/530)	K_3 (690/735)	K_4 (460/690)	K_5 (430/690)	K_6 (530/690)	K_7 (430/735)	K_8 (460/735)	K_9 (530/735)	K_{10} (430/530)
Rye "Belta"	1.1	1.6	2.8	1.3	1.3	0.8	3.8	3.6	2.3	1.7
Wheat soft "Belozerkovskaya-47"	1.0	1.1	2.9	1.1	1.1	1.0	3.1	3.2	2.9	1.1
Wheat winter soft "Ivanovskaya-60"	0.9	1.2	3.0	1.1	0.9	0.9	2.8	3.1	2.6	1.1
	1.0	1.4	2.5	6.4	6.2	4.5	15.2	15.9	11.0	1.4
	0.9	1.4	6.3	3.5	3.2	2.5	19.9	22.1	15.8	1.3
	1.0	1.3	2.8	2.4	2.4	1.9	6.7	6.7	5.3	1.1
Wheat winter soft "Mironovskaya-808"	1.0	1.1	3.4	1.0	1.0	0.9	3.3	3.4	3.1	1.1
"Kharkovskayaa-46"	0.9	1.5	3.4	1.2	1.1	0.8	3.6	4.0	2.7	1.3

(Table Contd.)

(Table Contd.)

Wheat winter stiff	1.0	1.4	3.7	1.2	1.2	0.9	4.5	4.5	3.3	1.4
"Svetlana"	1.0	1.9	4.0	0.2	1.3	0.7	5.4	2.9	2.9	1.9
	1.0	1.2	3.5	1.4	1.4	1.1	4.8	4.9	4.0	1.2
	1.0	1.3	5.3	2.7	2.4	2.1	12.5	14.1	11.0	1.1
	1.0	1.3	4.1	0.9	0.9	0.7	3.6	3.7	2.8	1.3
	1.0	1.2	4.1	1.5	1.6	1.2	6.3	6.1	5.0	1.3
	0.9	1.2	4.5	1.5	1.4	1.2	6.1	6.6	5.4	1.1
	1.0	1.2	1.0	1.9	1.8	1.7	1.8	1.9	1.7	1.1
	0.9	1.0	3.5	0.6	0.6	0.6	2.0	2.2	2.1	0.9
	1.0	1.2	3.5	0.8	0.8	0.6	2.6	2.7	2.2	1.2
Wheat spring stiff "Kharkov-7"	1.0	1.2	3.5	1.0	0.9	0.8	3.2	3.3	2.9	1.1
Wheat spring soft "Diana"	0.9	1.2	3.9	0.8	0.8	0.7	2.9	3.1	2.6	1.1
Mean values	1.0	1.3	3.5	1.6	1.4	1.1	3.9	3.9	3.1	1.3

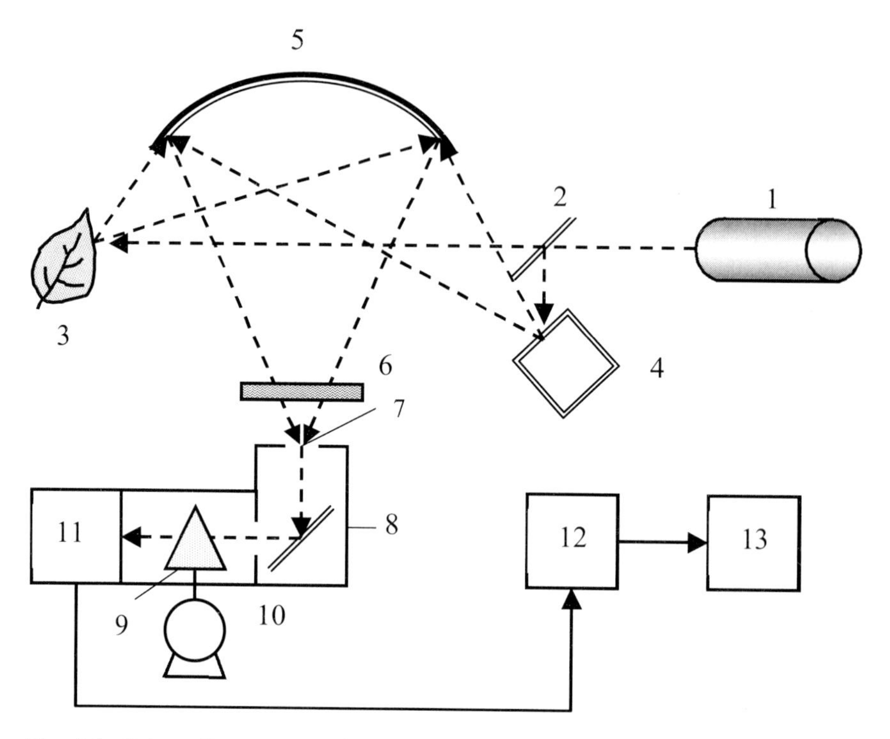

Fig. 8.1 Schematic representation of laser spectrofluorometer (explanation in text) [Posudin, 1998]

Wheat, rye, barley and hybrids have the following values: $\mathcal{K}_4, \mathcal{K}_5 \approx$ 1–8, $\mathcal{K}_6 \approx$ 1–6, $\mathcal{K}_7, \mathcal{K}_8 \approx$ 1–13, $\mathcal{K}_9 \approx$ 3–9, while the value of corn is one order higher than these ratios which can reach values of 11–27.

It was interesting to study these ratios for the different species of the same family (wheat). The results (Table 8.2) demonstrate the narrow interval of the values of such fluorescence indices as $\mathcal{K}_2 \approx$ 1.4–1.9, $\mathcal{K}_3 \approx$ 3.3–4.6, and a wider interval for $\mathcal{K}_4 \approx$ 0.8–5.9.

The growth of the plant from sprout to yellow stem is characterized by the decrease of mean values of fluorescence indices such as $\mathcal{K}_4 \approx 1.6$, $\mathcal{K}_5 \approx 1.4$, $\mathcal{K}_6 \approx 1.1$, $\mathcal{K}_7$, $\mathcal{K}_8 \approx 4$, $\mathcal{K}_9 \approx 3$, while the values of $\mathcal{K}_1 \approx 1$ and $\mathcal{K}_3 \approx$ 3–4 remained constant (Table 8.3).

The dependence of these ratios on the length of stems (for wheat winter soft "Ivanovskaya-60" and wheat spring stiff "Svetlana") is also demonstrated in Table 8.3. The values of fluorescence indices $\mathcal{K}_1$, $\mathcal{K}_2$, $\mathcal{K}_3$, $\mathcal{K}_{10}$ did not depend on the length of the stem, while the values of $\mathcal{K}_4$ and $\mathcal{K}_9$ demonstrated a wider interval.

In such a way, the fluorescence indices K_4 = $F(450)/F(690)$, K_8 = $F(450)/F(735)$, K_3 = $F(690)/F(735)$, K_2 = $F(450)/F(530)$ are decreasing from etiolated to the green leaves. Such modifications of fluorescence spectra during greening of the leaves can be explained with difference of absorption spectra of etiolated and green leaves [Stober and Lichtenthaler, 1993]: the etiolated leaf demonstrates the broad absorption band of carotenoids near 425, 450 and 485 nm, while the green leaf exhibits an absorption maximum at 430 nm, a broad shoulder near 470–500 nm and a maximum at 676 nm.

The first stage of the dynamics of plant growth is characterized by the formation of etiolated sprouts which have a whitish-yellow colour at the top, bright-yellow in the middle part, and bluish-white at the base of the sprout. The etiolated (one to three days) sprouts demonstrate the intensive fluorescence in the blue part of the spectrum and weak fluorescence in the red part. The germinated seeds and etiolated sprouts have the identical shape of spectra; the seed spectra has a non-resolved blue band and a smaller red band. The etiolated leaves demonstrate the more intense blue fluorescence and the weakest red fluorescence at 690 nm. The transformation of the etiolated sprout to yellow stem is accompanied by the decrease of the intensity of the blue-green fluorescence, increase of shoulder at 530 nm, increase in the intensity of red band and appearance of maximum at 735 nm; the most prolonged illumination induces the decrease of red fluorescence (Fig. 8.2).

The presence of only one maximum at 690 nm in fluorescence emission spectrum of etiolated leaves can be explained by a low (trace) amounts of chlorophyllide and/or chlorophyll *a*; the decrease in the intensity of fluorescence in the blue part of the spectrum can be provoked by a partial re-absorption of the emitted fluorescence by the photosynthetic pigments such as chlorophylls *a* and *b*, and carotenoids; the disappearance of fluorescence in the green part of the spectrum can be explained by the re-absorption effects of chlorophyll *b* and carotenoids [Lichtenthaler et al., 1981; Lichtenthaler, 1987; Stober and Lichtenthaler, 1992]. Appearance of maximum at 735 nm in the fluorescence spectrum of greening sprouts means that far-red chlorophyll fluorescence is slightly affected by the re-absorption processes [Stober and Lichtenthaler, 1993].

The same results were obtained by Stober et al. (1994). The blue-green fluorescence is a typical signature of yellowish etiolated leaves, white chlorophyll, and carotenoid-free white leaves. These leaves exhibited blue fluorescence 1.5 times stronger than the green leaf, and a distinct shoulder at 530 nm after illumination for two minutes with white light. The treatment of the leaves with the herbicide nonfluorazone (SAN

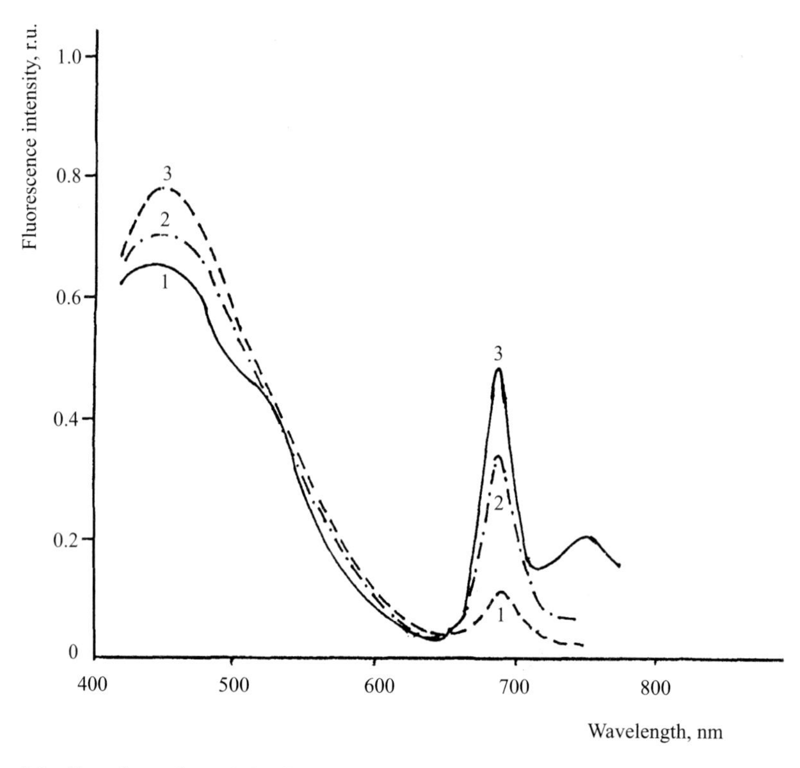

Fig. 8.2 Transformation of the fluorescence emission spectra from etiolated sprout to yellow stem: 1 – 1 hour, 2 – 5 hours, and 3 – 36 hours of illumination with white light [Ivanitskaya and Posudin, 1989]

9789), which blocked the synthesis and accumulation of the carotenoids in wheat and bleached chlorophyll, leads to a fourfold increase of the blue fluorescence as compared to green leaves. A similar effect was obtained after acetone-washing of the leaves and removing of chlorophyll and carotenoids.

The dependence of fluorescence properties of the leaves on the phase of growth can be explained in the following way. It is known that the biosynthesis of chlorophyll *a* in the leaves of the higher plants presents the complex process, which consists of the chain of photochemical transformations of chlorophyll from protochlorophyll. There are several dark and light stages of this process [Litvin and Belyaeva, 1971] which are presented in Fig. 8.3. The initial pigment form, protochlorophyll, participates in the photochemical reactions and undergoes a number of transformations, which are accompanied by a change of spectral properties (here the upper index corresponds to the fluorescence

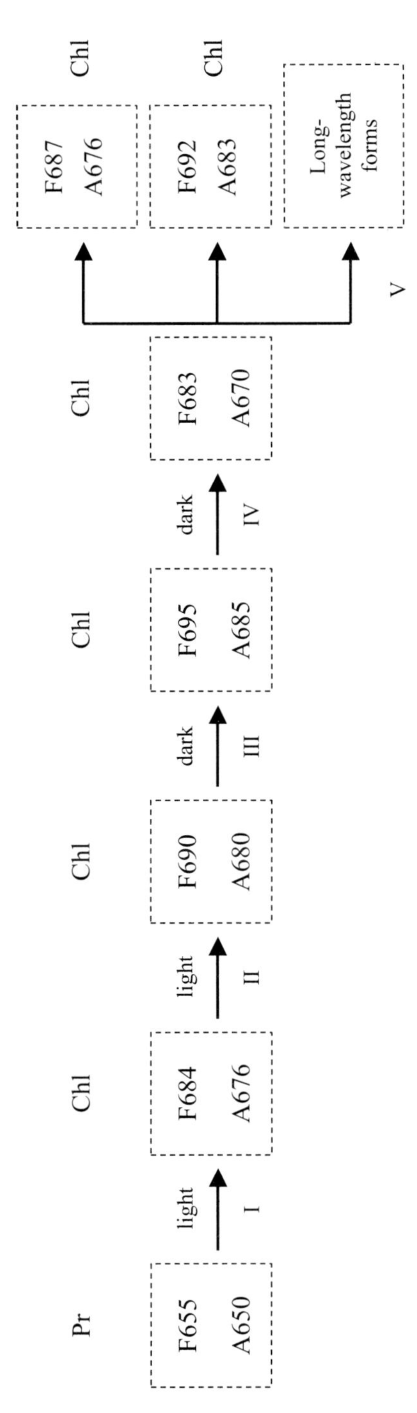

Fig. 8.3 Biosynthesis of chlorophyll *a* in the leaves of the higher plants as the complex process which consists of the chain of photochemical transformations of chlorophyll from protochlorophyll [Litvin and Belyaeva, 1971]

maximum, and the lower one to the absorption maximum). Hence, the first stage is the fast photoreaction of hydration and dark reorganization of pigment-protein complex of aggregated chlorophyll; the second stage presents the formation of phytol molecules and desegregation of pigment forms; the third stage is characterized by the accumulation of chlorophyll and increase in pigment concentration; the fourth stage is related to the appearance of the majority of pigment forms typical of the green leaf; and finally, the fifth stage demonstrates the consequent formation and accumulation of the long-wavelength forms of chlorophyll. All these stages are accompanied by corresponding spectral changes: the increasing intensities of red and far-red bands, and spectral shifts from 655 to 683–692 nm and to 722–738 nm.

The fluorescence intensity depends on the part of sprout; the intensity of blue and red bands is minimal for the middle segment of the leaf and maximal for the upper and lower segments (Figs. 8.4–8.6).

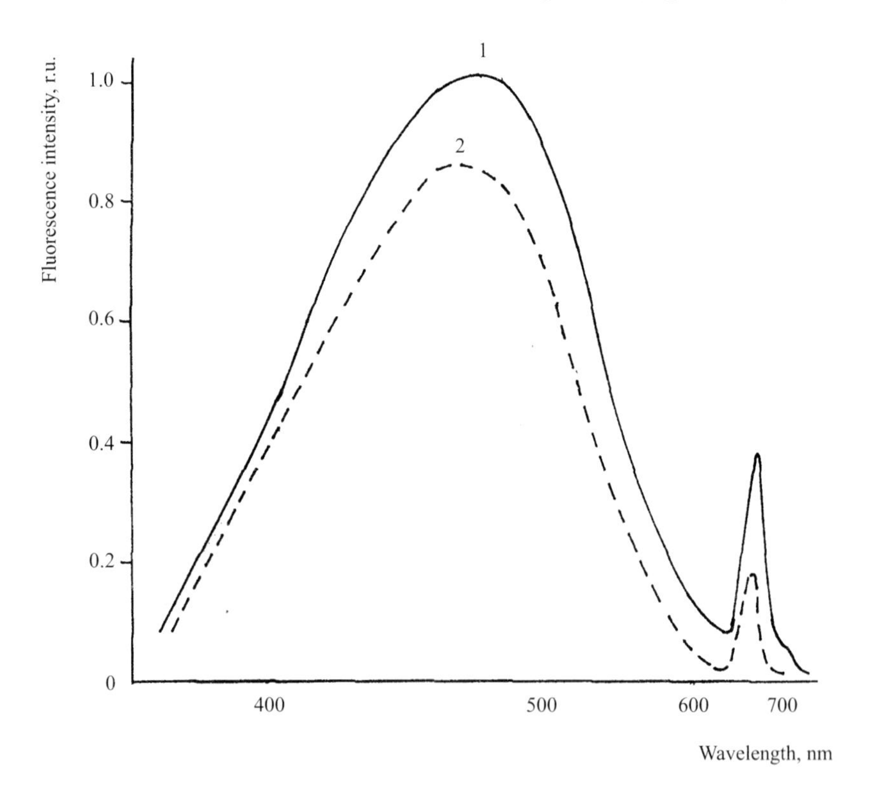

Fig. 8.4 Dependence of fluorescence intensity of winter stiff wheat "Korund" on parts of the leaf: 1 – lower segment; 2 – middle segment [Ivanitskaya and Posudin, 1989]

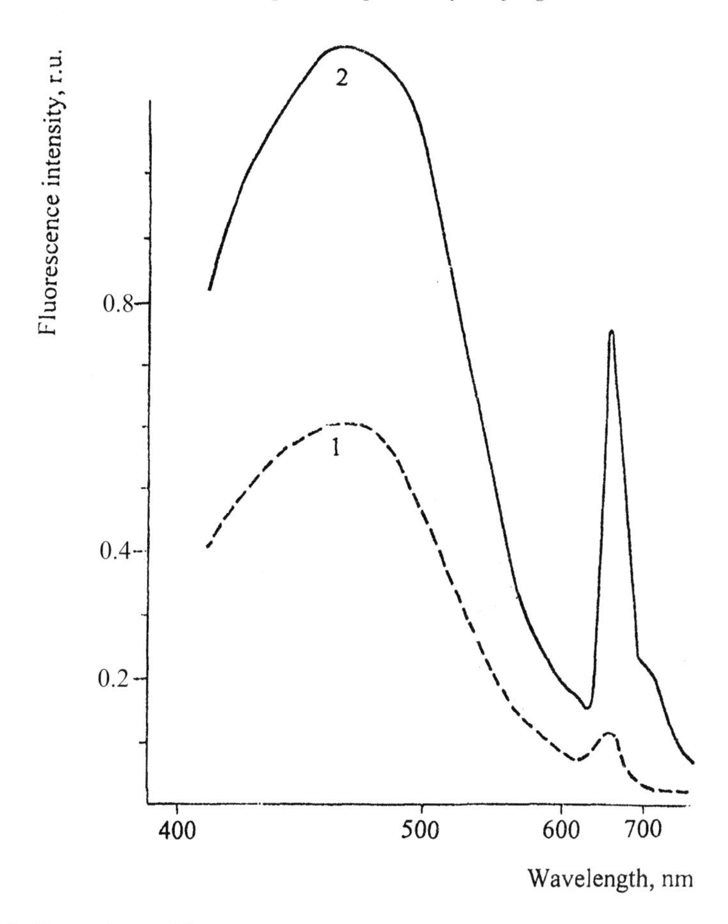

Fig. 8.5 Dependence of fluorescence intensity of rye "Zhitomyrskaya-tetra" on parts of the leaf: 1 – upper segment; 2 – middle segment [Ivanitskaya and Posudin, 1989]

8.3.3 Effect of Pre-illumination

The effect of pre-illumination of the etiolated stems of winter soft wheat ("Ivanovskaya-60" and "Mironovskaya-808") by white light (4000 lx during five hours) is presented in Table 8.4. There is a certain decrease in the fluorescence indices such as K_3 - K_4.

Longer illumination (during 36 hours) of spring wheat provokes the greening of stems; the fluorescence indices K_3 - K_4 decrease (Fig. 8.7) after irradiation (Table 8.5).

Such a behaviour of the fluorescence spectra of etiolated leaves during greening correlates with the total chlorophyll content and can be explained by partial re-absorption of the emitted fluorescence by the

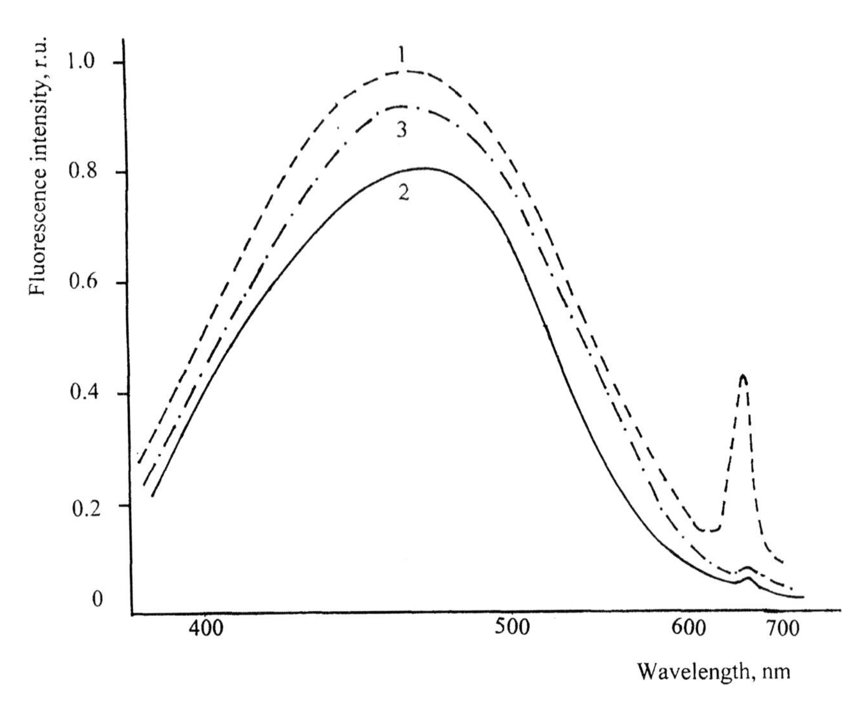

Fig. 8.6 Dependence of fluorescence intensity of hybrid amphidiploid "Diploid 3/5" on parts of the leaf: 1 – upper segment; 2 – middle segment; 3 – lower segment [Ivanitskaya and Posudin, 1989]

photosynthetic pigments [Stober and Lichtenthaler, 1992; Lichtenthaler et al., 1993].

8.3.4 Fluorescence Emission Spectra of Vegetables

It is very informative to find the specific spectral features of different plant species; such spectral "fingerprints" can be used in the remote sensing vegetation for identification and inventory of agricultural fields.

The fluorescence characteristics of leaves of the following vegetables have been investigated: potato, cucumber, tomato, bean, vegetable marrow, pumpkin, pepper, haricot and horse-radish (Table 8.6).

The most prominent difference in the fluorescence spectra is obtained accurately for vegetables; the species demonstrate the different fluorescence intensities in blue, red and far-red parts of the spectrum (Figs. 8.8-8.16).

Table 8.4 Fluorescence indices of yellow stems of winter soft wheat before and after irradiation with white light during five hours

Species	*Before Irradiation*			*After Irradiation*		
	K_2 *(460/530)*	K_3 *(690/735)*	K_4 *(460/690)*	K_2 *(460/530)*	K_3 *(690/735)*	K_4 *(460/690)*
"Ivanovskaya-60"	1.3	3.6	2.7	1.4	2.7	1.4
	1.2	3.8	1.2	1.5	1.4	1.4
	1.5	3.4	7.2	1.3	2.6	1.5
	1.4	4.8	4.1	1.5	2.3	5.0
	1.2	4.1	2.0	1.3	1.8	0.8
	1.3	2.8	2.4	1.3	2.0	1.0
	1.2	3.0	1.1	1.3	2.2	1.3
	1.4	2.5	6.4	1.4	0.8	1.2
	1.4	6.3	3.5			
"Mironovskaya-808"	1.1	4.4	1.1	1.2	2.2	1.4
	1.3	4.5	1.5	1.2	2.1	3.0
	1.1	3.4	1.0	1.2	1.9	2.1
				1.5	1.9	1.4
				1.5	1.9	1.2
				1.2	1.9	1.2
				1.2	2.2	2.6
				1.2	1.9	1.8

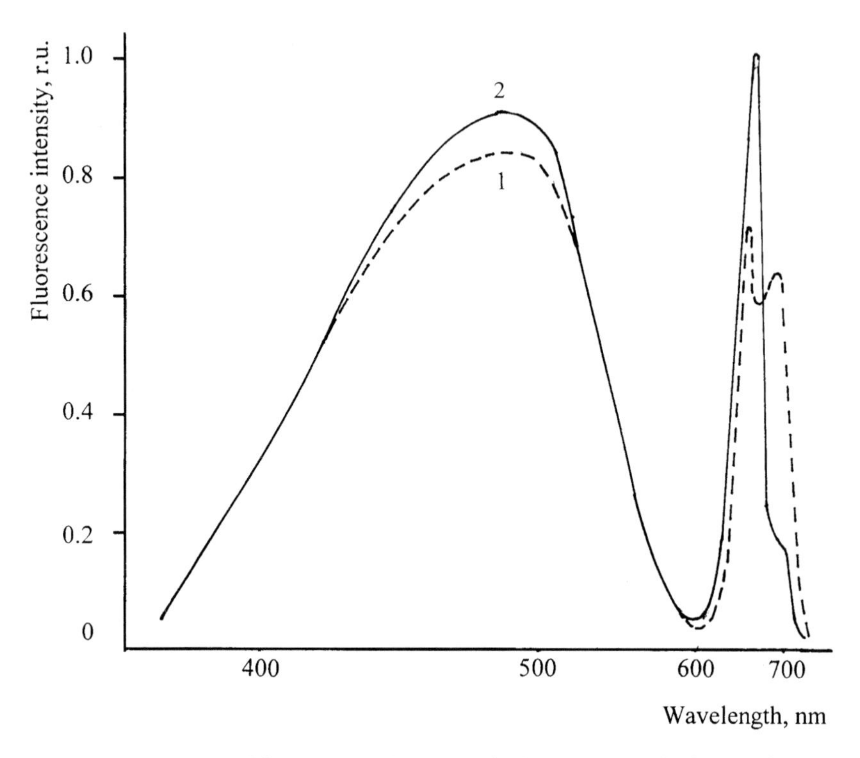

Fig. 8.7 Dependence of fluorescence intensity of wheat winter stiff “Svetlana” on the illumination of the sample during 36 hours with white light (4,000 lx): 1 – before, 2 – after illumination [Ivanitskaya and Posudin, 1989]

8.3.5 Effect of the Leaf Side

An overwhelming majority of the agronomic plants have bifacial leaves: the upper (*ventral*) side of the leaf includes palisade parenchyma with densely packed cells; the lower (*dorsal*) side contains spongy parenchyma with less quantity of cells. Such leaf morphology is typical for C_3-plants. In that way the upper side of the leaf presents a higher concentration of chlorophyll than the lower one. It was shown [Lichtenthaler and Rinderle, 1988] that when the fluorescence is excited and sensed from the upper leaf side, the maxima of emission at 690 and 735 nm are approximately equal in height; the value of the maximum at 690 nm is higher than the value of the maximum at 735 nm during excitation of fluorescence from the lower side of the leaf.

The C_4-plants are characterized by the so-called *Kranz-anatomy* – the concentric disposition of mesophyll cell layers; the difference in the

Table 8.5 Fluorescence indices of yellow stems of spring wheat after irradiation with white light during 36 hours

Cereal Species	*Fluorescence Indices*									
	K_1 *(430/460)*	K_2 *(460/530)*	K_3 *(690/735)*	K_4 *(460/690)*	K_5 *(430/690)*	K_6 *(530/690)*	K_7 *(430/735)*	K_8 *(460/735)*	K_9 *(530/735)*	K_{10} *(430/530)*
"Diana"	0.9	1.2	0.9	1.0	0.9	0.9	0.8	0.9	0.8	1.0
"Vector"	0.9	1.3	0.9	0.8	0.8	0.6	0.7	0.8	0.6	1.2
	1.2	1.2	1.0	0.7	0.9	0.6	0.8	0.7	0.6	1.5
	0.9	1.7	1.1	1.8	1.6	1.1	1.8	2.0	1.2	1.5
	1.0	1.4	1.2	1.0	0.9	0.7	1.1	1.1	1.0	1.4
	0.9	1.2	1.2	0.9	0.8	0.8	1.0	1.1	0.9	1.0
	1.0	1.4	0.7	1.4	1.3	1.0	1.0	1.0	0.7	1.3
	1.0	1.5	1.6	1.3	1.3	0.9	2.1	2.1	1.5	1.4
"Kharkov-7"	0.9	1.3	0.9	0.9	0.8	0.7	0.7	0.8	0.6	1.1
	1.0	1.5	0.9	1.2	1.2	0.8	1.1	1.1	0.8	1.4
	0.9	1.3	1.0	1.4	1.2	1.1	1.2	1.3	1.1	1.1
"Svetlana"	1.0	1.6	0.8	0.8	0.8	0.5	0.6	0.7	0.4	1.5
	0.9	1.2	1.7	1.2	1.1	1.0	1.0	2.1	1.7	1.2
	1.0	1.5	1.0	1.5	1.6	1.1	1.6	1.5	1.1	1.5
Mean values	1.0	1.4	1.1	1.1	1.1	0.8	1.2	1.2	0.9	1.3

Table 8.6 Fluorescence indices of green leaves of vegetables

Vegetable Species	*Fluorescence Indices*								
	K_1 (430/460)	K_2 (460/530)	K_3 (690/735)	K_4 (460/690)	K_5 (430/690)	K_6 (530/690)	K_7 (430/735)	K_8 (460/735)	K_9 (530/735)
Cucumber	0.7	0.6	1.3	0.9	0.6	0.2	0.1	1.1	0.2
	0.8	0.6	1.2	0.1	0.1	0.2	0.1	0.1	0.2
Vegetable marrow	0.7	1.0	1.1	0.5	0.3	0.5	0.3	0.6	0.5
	0.7	1.1	1.4	0.6	0.4	0.5	0.5	0.7	0.6
	0.7	1.0	1.4	0.6	0.4	0.6	0.5	0.8	0.8
Tomato	0.6	0.9	1.1	1.6	1.0	1.8	1.1	1.7	1.9
	0.7	0.9	0.9	2.6	1.7	2.9	1.6	2.4	2.7
	0.6	1.7	1.0	1.9	1.2	2.3	1.3	2.0	2.3
Haricot	0.7	0.8	1.0	0.2	0.1	0.3	0.1	0.2	0.3
	0.7	1.0	1.0	0.3	0.2	0.3	0.2	0.3	0.3
	0.6	1.0	0.9	0.4	0.3	0.4	0.2	0.4	0.4
Sorrel	0.7	1.1	1.1	0.7	0.5	0.6	0.5	0.7	0.7
	0.7	1.1	1.2	0.8	0.6	0.8	0.8	1.0	0.9
Horse-radish	0.9	0.8	1.3	0.1	0.1	0.1	0.1	0.1	0.2
	0.5	0.2	1.5	0.1	0.1	0.7	0.1	0.2	1.1
	0.8	1.0	1.4	0.1	0.1	0.1	0.1	0.1	0.1
Currants	0.9	0.8	1.0	0.2	0.2	0.2	0.2	0.2	0.2
	1.1	1.0	0.8	0.2	0.2	0.2	0.1	0.1	0.1
	0.9	0.9	1.8	0.2	0.2	0.3	0.2	0.2	0.3
Raspberry-cane	0.6	0.6	0.9	1.0	0.6	1.7	0.5	0.8	1.5
	0.7	0.9	1.0	1.0	0.7	0.5	0.7	1.0	1.1
	0.8	0.8	1.0	0.8	0.6	1.0	0.7	0.8	1.0

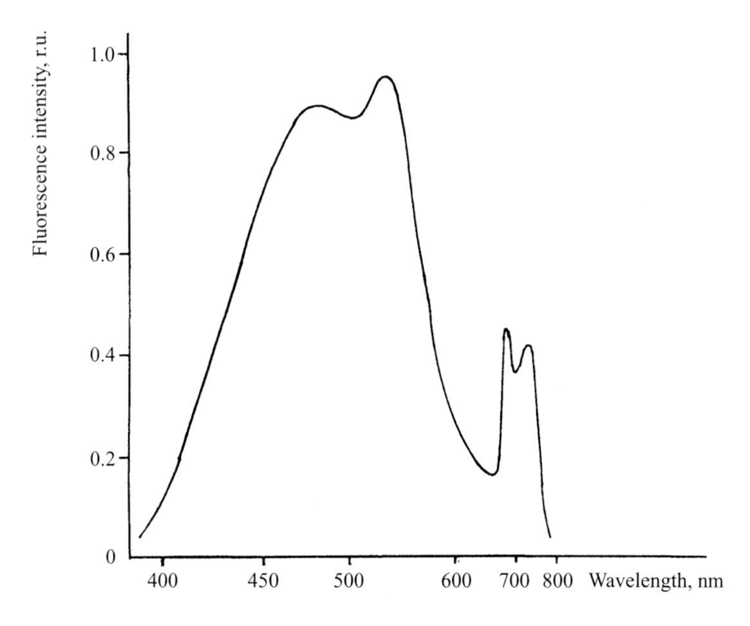

Fig. 8.8 Fluorescence emission spectrum of tomato [Ivanitskaya and Posudin, 1989]

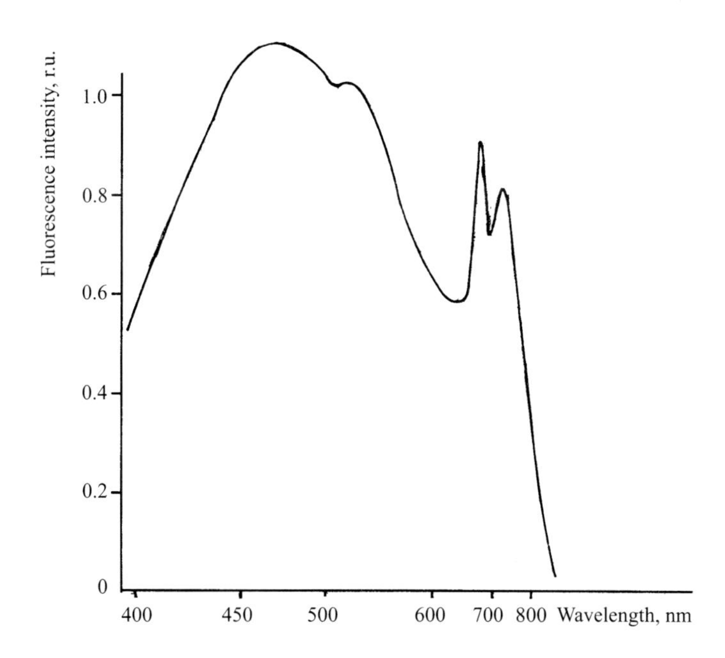

Fig. 8.9 Fluorescence emission spectrum of potato [Ivanitskaya and Posudin, 1989]

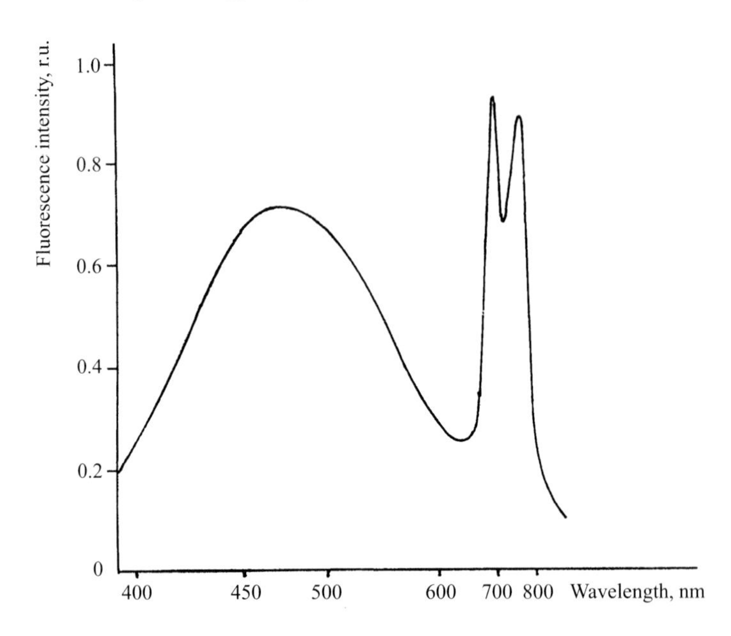

Fig. 8.10 Fluorescence emission spectrum of pumpkin [Ivanitskaya and Posudin, 1989]

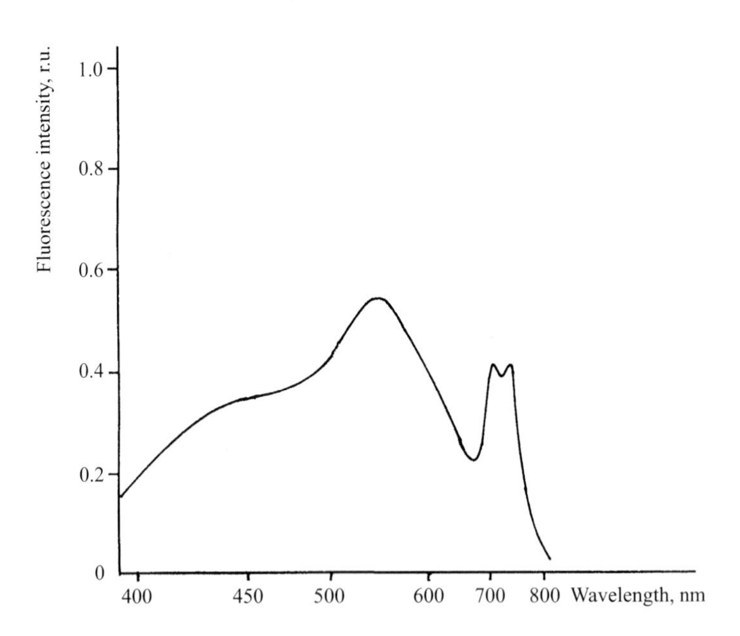

Fig. 8.11 Fluorescence emission spectrum of pepper [Ivanitskaya and Posudin, 1989]

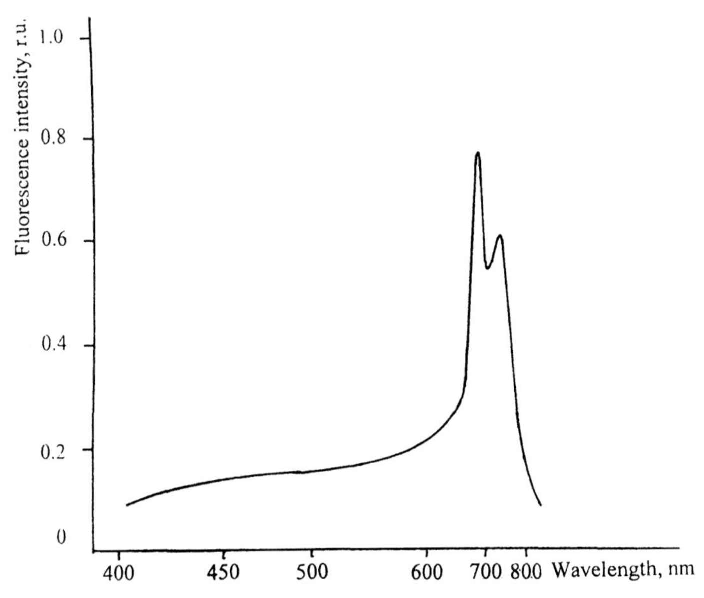

Fig. 8.12 Fluorescence emission spectrum of horse-radish [Ivanitskaya and Posudin, 1989]

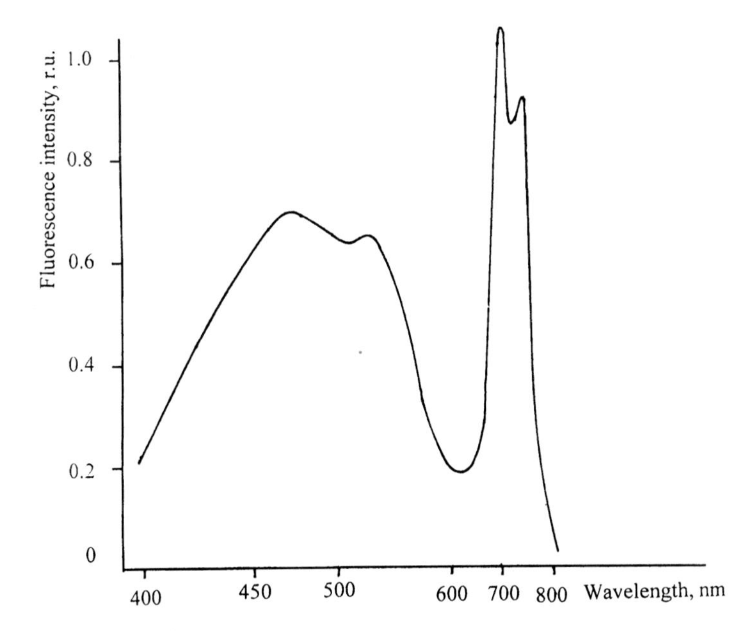

Fig. 8.13 Fluorescence emission spectrum of sorrel [Ivanitskaya and Posudin, 1989]

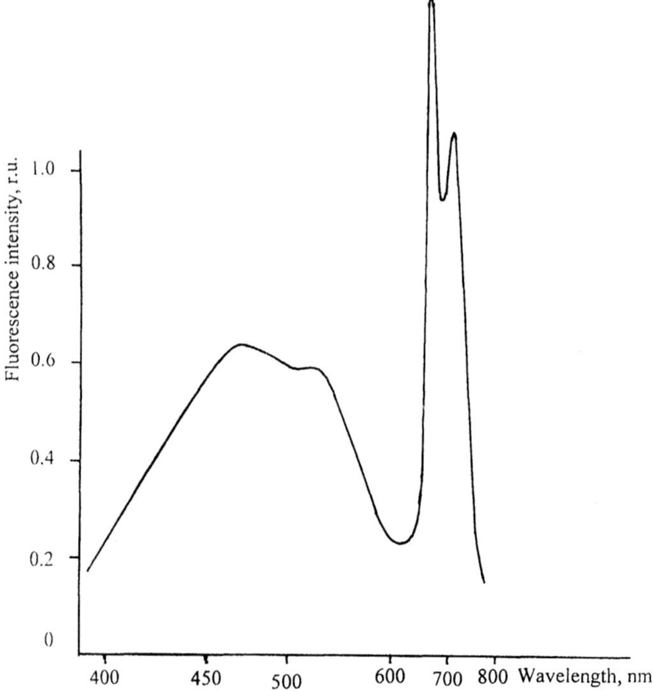

Fig. 8.14 Fluorescence emission spectrum of vegetable marrow [Ivanitskaya and Posudin, 1989]

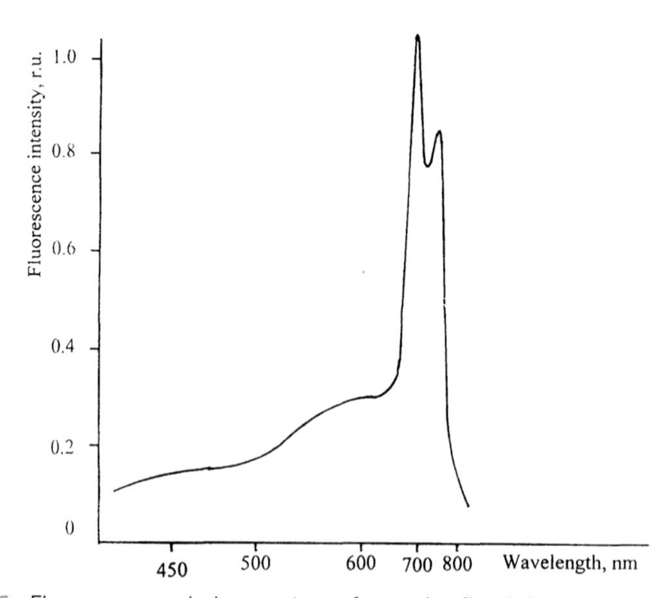

Fig. 8.15 Fluorescence emission spectrum of cucumber [Ivanitskaya and Posudin, 1989]

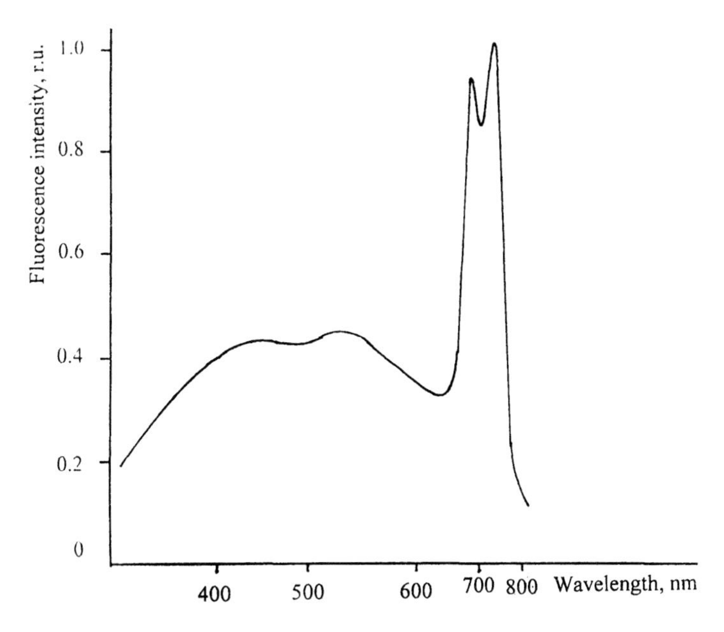

Fig. 8.16 Fluorescence emission spectrum of haricot [Ivanitskaya and Posudin, 1989]

fluorescence spectra of such equifacial leaves (for example, the leaves of maize) between the upper and lower side is rather small. The authors consider that if the concentration of chlorophyll is higher in the upper side, the probability of re-absorption of the emitted fluorescence is also higher; correspondingly, the values of the ratio *F690/F735* are: 0.85–1.10 for upper side and 1.20–1.70 for the lower side. Chappelle et al. (1984*a*, *b*) established that laser-induced fluorescence of the back side of the soybean leaf is greater than that of the front side. These fluorescence distinctions are related to the differences either in morphological structure, or in the photosynthetic properties of both sides.

The difference in the fluorescence spectra between upper and lower leaf sides is also demonstrated by several C_4 plants such as maize (Fig. 8.17).

8.3.6 Effect of Leaf Nodal Position and Age

The effect of nodal position of leaves of the vegetables is presented in Table 8.7. This effect of the leaf position on the plant stem on the fluorescence characteristics can be related to the age of the leaf (the youngest dark green leaves are located in the upper nodes) and to the different levels of photosynthetic activity. The older, yellow-green leaves

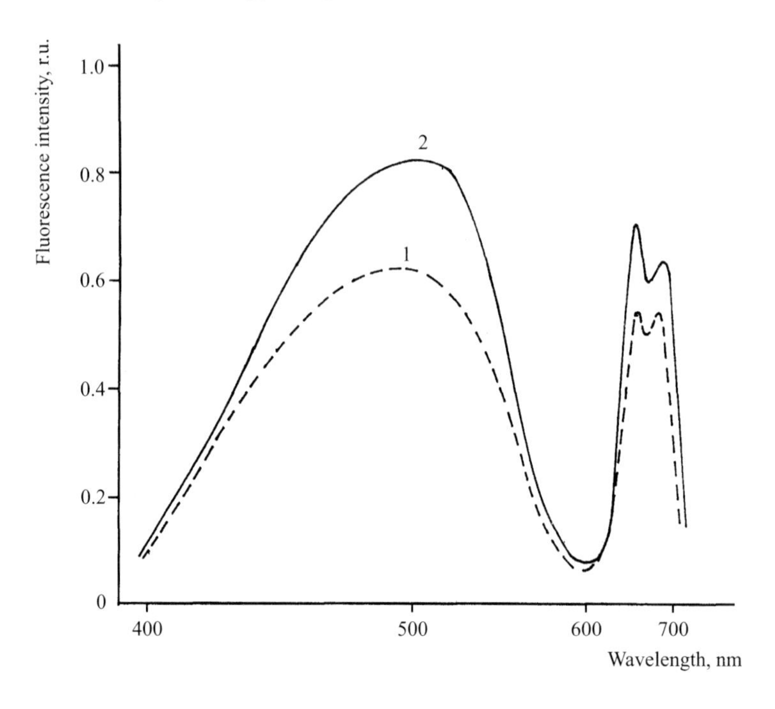

Fig. 8.17 Fluorescence emission spectra of upper (1) and lower (2) sides of maize – representative of C_4 plants [Ivanitskaya and Posudin, 1989]

of potato demonstrate lower intensities of fluorescence than the younger dark green leaves (Fig. 8.18). These results are in complete agreement with those observed by Lang et al. (1992), i.e. decrease of the blue fluorescence emission from young to an old leaf. The reason is the same - different concentrations of chlorophyll during leaf development and correspondingly different levels of re-absorption of fluorescence.

8.3.7 Effect of Fertilization

The effect of different natural and artificial fertilizers, such as manure, $N_{30}P_{30}K_{30}$ and $N_{15}P_{15}K_{15}$, and "biohumus" (product of worm activity in the soil), just as the combinations of these fertilizers, on the fluorescence of tomato leaves (sort "Kechkemeti 549") was observed by the author and his colleagues [Gorodny et al., 1992]. It was shown that the fluorescence parameters could be used for estimation of fertilization efficiency.

8.3.8 Effect of Plant Diseases

It was very interesting to establish the effect of plant pathology on the chlorophyll fluorescence of leaves. Bean plants with affected root system [Posudin et al., 1992] have been used in this experiment. The level of the

Table 8.7 Fluorescence indices of different nodal positions of vegetable leaves

Vegetable Species	*Fluorescence Indices*								
	K_1 *(430/460)*	K_2 *(460/530)*	K_3 *(690/735)*	K_4 *(460/690)*	K_5 *(430/690)*	K_6 *(530/690)*	K_7 *(430/735)*	K_8 *(460/735)*	K_9 *(530/735)*
Cucumber (Upper)	0.9	0.8	1.2	0.1	0.1	0.1	0.1	0.1	0.1
	0.8	0.6	1.2	0.1	0.1	0.2	0.1	0.1	0.3
	0.7	0.7	1.4	0.1	0.1	0.1	0.1	0.1	0.2
Cucumber (Lower)	0.8	0.9	1.9	0.4	0.8	0.5	0.6	0.8	0.9
	0.8	0.9	1.3	0.1	0.5	0.1	0.6	0.8	0.9
	0.8	0.9	1.4	0.5	0.5	0.7	0.6	0.8	1.0
Cucumber (Middle)	0.7	0.6	1.3	0.7	0.1	0.3	0.1	0.2	0.3
	0.7	0.8	1.2	0.2	0.1	0.2	0.2	0.2	0.3
	0.7	0.7	1.2	0.1	0.1	0.1	0.1	0.2	0.2
Vegetable marrow (Upper)	0.7	1.2	1.5	1.1	0.8	0.9	1.2	1.7	1.4
	1.2	1.3	1.4	0.5	0.6	0.4	0.8	0.7	0.5
	0.1	1.2	1.5	0.3	0.2	0.3	0.3	0.5	0.4
Vegetable marrow (Lower)	0.7	1.0	1.2	1.1	0.7	1.0	0.8	1.3	1.2
	0.7	1.1	1.2	1.2	0.9	1.1	1.1	1.5	1.3
	0.7	1.0	1.5	0.8	0.6	0.8	0.9	1.3	1.3
Vegetable marrow (Middle)	0.6	0.9	1.4	1.6	1.0	1.7	1.4	2.2	2.4
	0.6	0.9	1.2	1.1	0.7	1.3	0.8	1.3	1.5
	0.8	1.3	1.3	2.1	1.7	1.6	2.1	1.7	2.0
Tomato (Upper)	0.6	1.0	1.0	1.4	0.9	1.4	0.9	1.4	1.5
	0.7	1.1	1.0	1.0	0.7	0.9	0.7	1.0	0.9
	0.6	1.0	0.8	1.0	0.6	0.9	0.7	1.1	1.1

(Table Contd.)

(Table Contd.)

Tomato (Lower)	0.6	1.0	1.2	3.2	1.2	2.0	1.4	2.5	2.4
	0.6	0.9	1.2	0.9	0.5	0.9	0.6	1.0	1.1
	0.7	0.9	1.2	0.9	0.5	0.9	0.6	1.0	1.1
Tomato (Middle)	0.7	1.1	1.3	1.4	1.0	1.3	1.2	1.8	1.7
	0.7	1.1	1.0	2.5	1.7	2.2	1.8	2.7	2.4
	0.6	1.1	0.4	3.3	2.0	3.1	0.8	1.3	1.2
Pumpkin (Upper)	0.7	1.2	1.4	0.2	0.2	0.2	0.2	0.3	0.2
	1.2	1.1	1.4	0.2	0.2	0.2	0.3	0.2	0.2
	0.7	1.1	1.4	0.2	0.1	0.2	0.2	0.4	0.2
	0.8	1.4	1.7	0.2	0.1	0.2	0.3	0.4	0.3
Pumpkin (Lower)	0.8	2.0	1.6	2.0	1.6	0.7	2.6	3.2	1.1
	0.7	1.3	1.1	1.1	0.8	0.9	0.9	1.3	1.0
	0.7	1.4	1.2	1.1	0.8	0.8	1.0	1.3	1.0
	0.7	1.3	1.0	1.4	1.0	1.1	1.0	1.4	1.1
Pumpkin (Middle)	0.7	1.1	1.2	0.7	0.5	0.7	0.6	0.8	–.8
	0.9	0.9	1.2	0.6	0.5	0.6	0.6	0.7	0.8
	0.8	1.3	1.4	0.5	0.4	0.4	0.6	0.8	0.6
	0.7	1.3	1.3	1.1	0.8	0.9	1.0	1.4	1.1
Potato (Upper)	0.7	1.1	1.2	1.3	1.0	1.2	1.2	1.6	1.5
	0.7	1.0	1.2	0.8	0.6	0.8	0.7	1.0	1.0
	0.7	1.0	1.3	0.8	0.5	0.8	0.7	1.0	1.0
Potato (Lower)	0.6	0.9	1.3	0.9	0.5	1.0	0.6	1.1	1.2
	0.6	1.0	1.3	1.1	0.7	1.1	0.9	1.4	1.4
	0.6	1.0	1.1	1.0	0.6	1.0	0.7	1.1	1.1

(Table Contd.)

(Table Contd.)

Potato (Middle)	0.6	0.7	1.2	0.7	0.6	0.9	0.8	1.1	1.1
	0.6	1.0	1.2	0.9	0.8	0.9	0.8	1.1	1.1
	0.6	1.0	1.3	0.6	0.3	0.6	0.4	0.8	0.7
	0.6	1.1	1.1	0.9	0.5	0.8	0.5	1.0	0.9
Sunflower (Upper)	0.6	1.1	1.0	0.4	0.2	0.4	0.2	0.4	0.4
	0.6	1.1	1.5	0.7	0.4	0.6	0.6	1.0	0.9
	0.6	1.1	1.3	0.8	0.5	0.7	0.6	1.1	1.0
	0.6	1.1	1.2	0.3	0.2	0.3	0.2	0.4	0.4
Sunflower (Lower)	0.5	0.7	1.2	1.4	0.8	2.0	0.9	1.7	2.3
	0.5	0.7	1.1	1.7	0.9	2.4	1.0	1.9	2.6
	0.5	0.8	1.3	1.6	0.8	2.2	1.1	2.2	2.9
Sunflower (Middle)	0.6	1.0	1.4	3.1	2.0	3.1	2.7	4.3	4.2
	0.6	1.1	1.3	3.3	2.1	3.1	2.8	4.3	4.1
Pepper (Upper)	0.7	0.7	1.1	0.3	0.2	0.4	0.2	0.3	0.4
	0.8	0.8	1.0	0.8	0.7	1.2	0.7	0.9	1.3
	0.7	1.0	0.7	2.1	1.6	1.9	1.1	1.4	1.3
Pepper (Lower)	0.7	0.9	0.9	0.7	0.5	0.8	0.4	0.6	0.6
	0.7	0.8	0.8	0.7	0.5	1.2	0.4	0.8	1.0
	0.5	0.8	0.9	0.9	0.4	1.2	0.4	0.8	1.0
	0.7	0.8	0.9	0.8	0.6	1.0	0.5	0.7	0.8
Pepper (Middle)	0.7	0.7	0.8	0.6	0.5	1.0	0.4	0.5	0.8
	0.9	0.7	1.1	0.9	0.7	1.3	0.8	1.0	1.5
	0.8	0.7	1.1	0.9	0.6	1.3	0.7	1.0	1.5

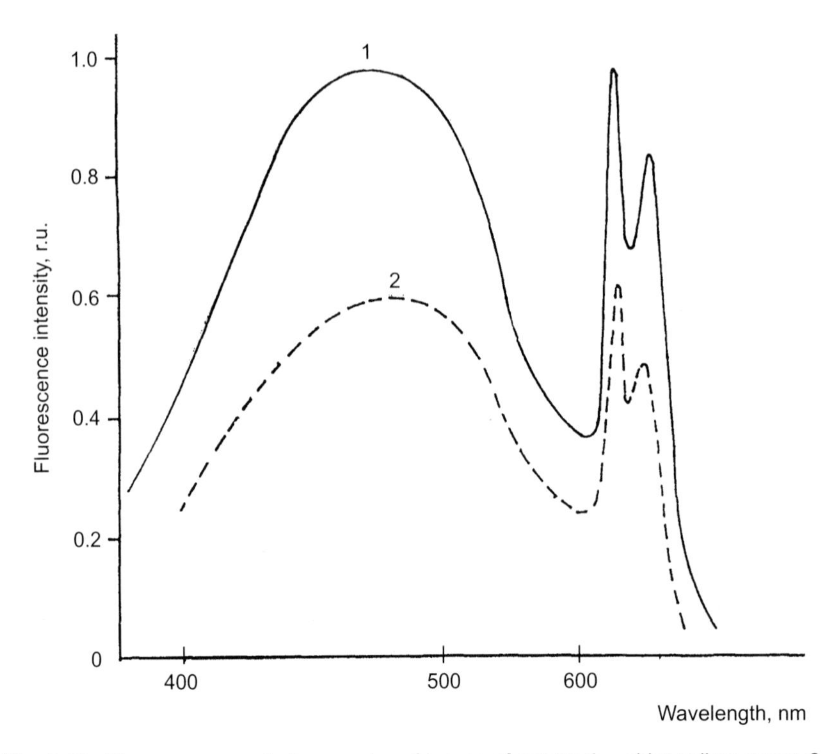

Fig. 8.18 Fluorescence emission spectra of leaves of potato: 1 – older yellow-green; 2 – younger dark green [Ivanitskaya and Posudin, 1989]

damage was estimated quantitatively due to 5-ball system (0 = healthy roots, 1 = small, 2 = middle, 3 = strong, 4 = utter defeat of the roots). The typical dependence of fluorescence intensity of the leaf on the level of the root defeat for different parts of leaf is presented in Fig. 8.19. Probably, fluorescence parameters can be applied as diagnostic criteria in the science of plant protection.

8.4 FLUORESCENCE INDUCTION KINETICS

8.4.1 Theory of Fluorescence Induction Kinetics

If a green photosynthetically active leaf is adapted to darkness during a certain period (15–20 minutes) of time, the cooperation of photosystems *PSI* and *PSII* is disturbed. The fact is that the illumination of the dark-adapted leaf provokes the so-called *fluorescence induction* (*Kautsky effect*) [Kautsky and Hirsch, 1931]. It is possible to distinguish two parts of the temporal behaviour of chlorophyll fluorescence: fast increase of

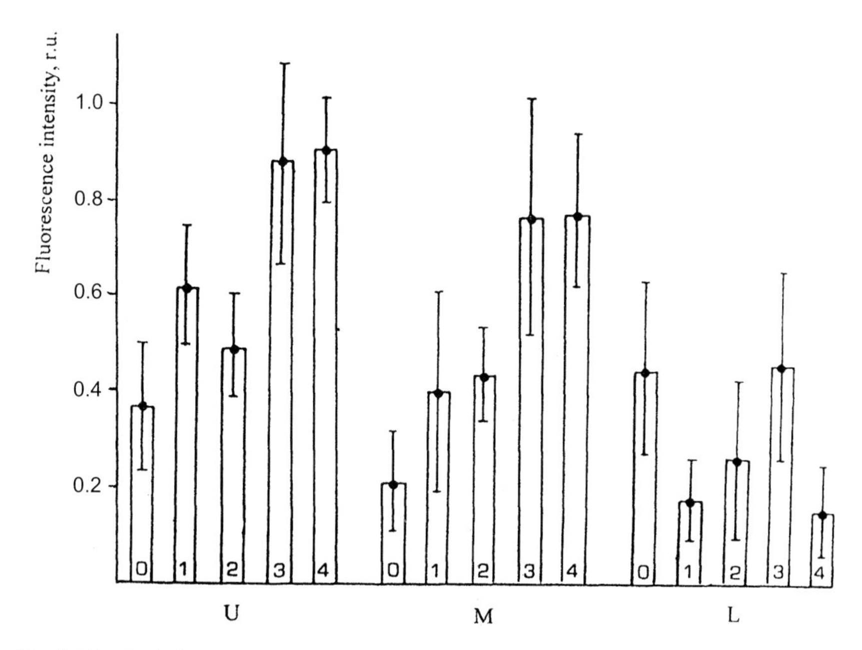

Fig. 8.19 Typical dependence of fluorescence intensity of the leaf on the level of the root damage which was estimated with different (from 0 to 4) points and from parts of leaf (U – upper, M – middle, L – lower) [Posudin et al., 1992]

fluorescence intensity up to maximal value f_m during 100–500 ms, and slow decrease of fluorescence up to stationary level f_s. The photosynthetic apparatus of the dark-adapted leaf is in state I, while the illumination induces its transfer to state II. A typical fluorescence induction curve is presented in Fig. 8.20.

8.4.2 Plant Samples

It is informative to provide the comparative analysis of fluorescence characteristics of different varieties of the same specie of the plants. The objective of this investigation is to study the changes in chlorophyll fluorescence during leaf development and under stress conditions of two varieties of green pea *Pisum sativum*, and to use chlorophyll fluorescence parameters as indices of plant status and taxonomic criteria. Two varieties of green pea *Pisum sativum*, *Determinantny* and *Bakara*, from the collection of National Agricultural University of Ukraine (Kiev), were used in these experiments. The plants were grown in the green-house of Botanical Garden of the University of Karlsruhe (Germany) at an ambient temperature of 22°C ± 4° of on a peat containing mineral nutrients (TKS II). Pieces of 9 mm diameter were cut off from the pea

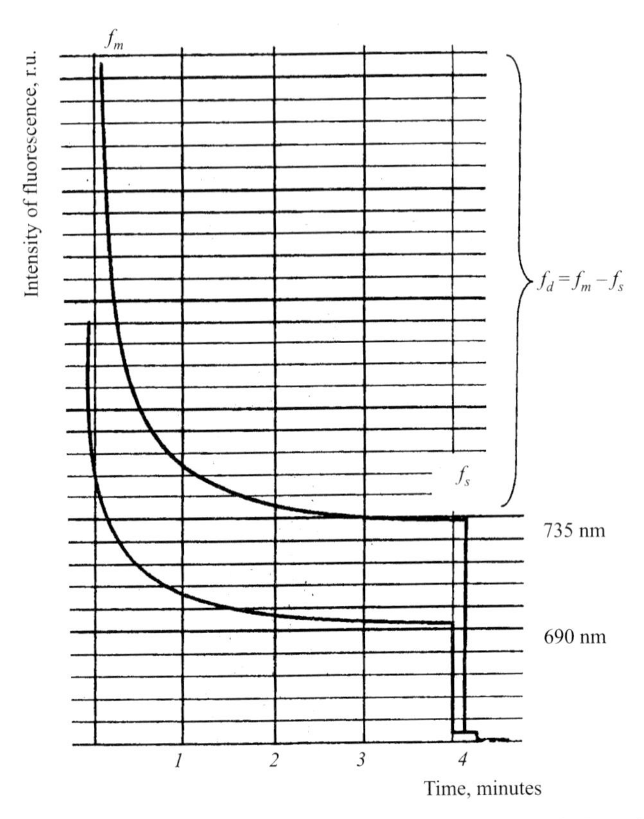

Fig. 8.20 Chlorophyll-fluorescence induction curves of 15-min predarkened leaves of green pea *Pisum sativum* recorded in the 690 nm region (lower curve) and 735 nm region (upper curve) [Posudin, 1997, 2002]

leaves and kept in darkness for 15 minutes before the chlorophyll fluorescence measurements.

8.5 INSTRUMENTATION

A principal set-up for measuring parameters of fluorescence induction is presented in Fig. 8.21. Two-wavelength fluorometer (National Agricultural University, Kiev, Ukraine) was used for illumination of the predarkened leaves and control of fluorescence emission at two wavelengths at 690 and 735 nm. It consists of mercurial lamp *1* DRK-120 of superhigh pressure which is used as the source of fluorescence excitation. The chlorophyll fluorescence kinetics of dark adapted pea leaves was analyzed during five minutes. The radiation of the lamp *1*

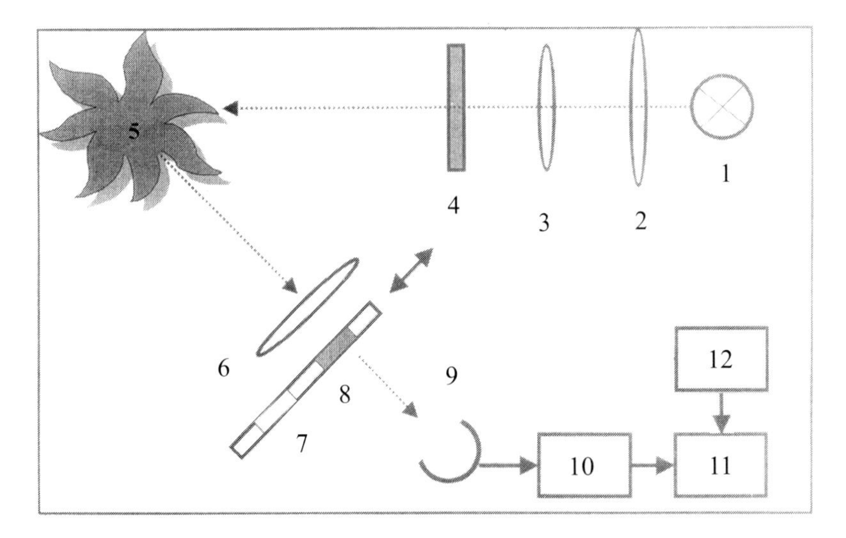

Fig. 8.21 Two-wavelength fluorometer (National Agricultural University, Kiev, Ukraine) used for illumination of the predarkened leaves and control of fluorescence emission at two wavelengths at 690 and 735 nm [Posudin and Kozhemjako, 2001; 2002]

passes through a condensor which is formed by two quartz lens *2* and *3*, excitation filter *4* (FS-1-2 and SZS-24-2) and enters the sample *5* - green leaf. The registering of fluorescence radiation of the sample is realized at an angle of 30° to the direction of excitation radiation in order to avoid the influence of it on the readout system. Lens *6* provides the focusing of fluorescence emission through filters *7* (690 nm) and *8* (735 nm) on the photomultiplier *9*. A signal from the photomultiplier enters the input of the amplifier *10* (F136) and recorder *11* (KSP-4). The photomultiplier, amplifier and recorder are linked with the source of voltage *12*.

8.6 RESULTS OF DETECTION OF CHLOROPHYLL FLUORESCENCE KINETICS

8.6.1 Fluorescence Indices

The following fluorescence indices were used for quantitative estimation of the changes of fluorescence parameters of the plant under stress conditions: maximum fluorescence f_m; steady-state fluorescence f_s; fluorescence decrease $f_d = f_m - f_s$; vitality index $Rfd = f_d/f_s$, which is measured in the 690- and 735-nm region: $Rfd(690)/Rfd(735)$; ratios f_m690/f_m735 and f_s690/f_s735, and the stress adaptation index $A_p = 1-[Rfd(735)+1]/[Rfd(690)+1]$.

8.6.2 Determination of Pigments

Photosynthetic pigments [chlorophyll ($a + b$) and total carotenoids ($x + c$)] were determined spectrophotometrically (UV2001 PC, Shimadzu, Duisburg, Germany) using the method of Prof. H.K. Lichtenthaler (1987). The pigments were extracted in 100% acetone and determined due to extinction coefficients at 750, 661.6, 644.8, and 470 nm.

The content of chlorophyll and carotenoids in the leaves of different levels of greening is presented in Table 8.8.

Table 8.8 Results of calculation of the pigment content (extract with 100% Acetone) in leaves of *Pisum sativum*

Variety of Plant	*Output Per Area*			
	Chl a [mg/m²]	*Chl b [mg/m²]*	*Carotenoids x+c [mg/m²]*	*Chlorophylls a+b [mg/m²]*
1. *Determinantny*, dark green, 1 node	248.05	83.82	63.63	331.87
2. *Determinantny*, dark green, 2 nodes	228.59	76.16	59.05	304.75
3. *Determinantny*, light green, 3 nodes	173.01	54.25	48.01	227.26
4. *Bakara*, dark green, 1 node	388.07	125.55	109.54	513.62

The dark green leaves (sample N1) of the type *Determinantny* contained a higher pigment content than the green leaves (sample N2) and light green leaves (sample N3). The dark green leaves of two kinds, *Determinantny* and *Bakara*, of the same age and nodal position showed considerable difference in pigment content (samples N1 and N4).

The parameters of chlorophyll fluorescence induction kinetics such as maximum fluorescence f_m, steady-state fluorescence f_s, fluorescence decrease $f_d = f_m - f_s$ and vitality index *Rfd'* (at 735 nm), *Rfd''* (at 690 nm), $f_m 690/f_m 735$ and $f_s 690/f_s 735$, and stress adaptation index A_p of the samples N1-N4 with different pigment content are presented in Table 8.9.

It is shown that the *Rfd* values measured in the 690 nm region were higher than in the 735 nm region. The values of ratios $f_m 690/f_m 735$ and $f_s 690/f_s 735$ decreased during greening of the leaves. Increase in the vitality indices *Rfd'* and *Rfd''* from light green to dark-green leaf was demonstrated.

Table 8.9 Dependence of the parameters of fluorescence induction kinetics of two varieties of leaves of *Pisum sativum* on the pigment content

Samples	735 *nm*			690 *nm*			*Indices*		
	f_d	f_s	*Rfd'*	f_d	f_s	*Rfd''*	$f_m(690)/f_m(735)$	$f_s(690)/f_s(735)$	A_p
Determinantny, dark green, 1 node	19.5	10.5	1.86	11.5	4.5	2.55	0.53	0.43	0.19
Determinantny, green, 2 nodes	19.5	10.7	1.82	12.0	5.0	2.40	0.57	0.48	0.17
Determinantny, light green, 3 nodes	20.0	11.2	1.78	13.0	5.5	2.36	0.63	0.58	0.17
Bakara, dark green, 1 node	19.5	10.0	1.95	10.5	4.0	2.62	0.49	0.40	0.18

Both the contents of the main plant pigments, chlorophyll and total carotenoids, increased during greening. In the leaves with normal photosynthetic apparatus, the main part of the absorbed light energy is used for photochemical work to drive photosynthesis, and a small portion (about 2–5%) of this energy is emitted as chlorophyll fluorescence. The dependence of the chlorophyll fluorescence parameters on the chlorophyll content can be explained in the following way [Lichtenthaler and Rinderle, 1988]: with increasing chlorophyll content, the short wavelength fluorescence becomes increasingly suppressed due to re-absorption of the emitted fluorescence by the chlorophyll which originates from the partial overlapping of the absorption and fluorescence emission spectra of the leaves between 640 and 740 nm. The influence of re-absorption of the emitted fluorescence is less in leaves with low chlorophyll content; at such low chlorophyll content the values of the main fluorescence parameters, such as f_d, f_s and *Rfd* exhibit higher values than in fully developed dark-green leaves. Besides, the changes in the pigment composition of the photosynthetic apparatus can also be taken into account. The difference of pigment content between two varieties of the same specie can be explained with the different levels of photosynthetic activity of these types.

8.6.3 Effect of Leaf Nodal Position and Age

The effect of nodal position of the leaves on chlorophyll fluorescence parameters is demonstrated in Table 8.10.

The values of fluorescence ratios f_m690/f_m735 and f_s690/f_s735, just as *Rfd* values declined with increasing leaf age from the youngest leaf of the first node to the older leaf (sixth node). The values of stress adaptation index A_p do not demonstrate the significant changes in the limits of errors of measurements.

There was a considerable difference between the fluorescence parameters such as f_m690/f_m735 and f_s690/f_s735 of two kinds of green peas of the same age and nodal position.

The effect of nodal position and correspondingly the age of the leaves on chlorophyll fluorescence parameters can be explained by the formation of photosynthetic apparatus during leaf development. There are a number of factors which can be taken into account in the course of plant ontogeny [Sestak, 1985; Sestak and Siffel, 1988]: anatomical parameters, size, number and ultrastructure of chloroplasts, contents of pigments and pigment-protein complexes, composition and activities of enzymes of the carbon fixation pathways, rates of gas exchange and biomass production. These factors change from the time of leaf

Table 8.10 Effect of nodal position and age of the leaves of two varieties of *Pisum sativum* on the fluorescence parameters

Nodal Position from	*735 nm*			*690 nm*			*Indices*		
Plant Apices	F_d	f_s	*Rfd'*	F_d	f_s	*Rfd"*	$f_m(690)/$ $f_m(735)$	$f_s(690)/$ $f_s(735)$	A_p
Determinantny, 6-node	19.5	10.8	1.89	12.3	5.0	2.46	0.57	0.46	0.16
oldest leaf	±0.4	±0.6	±0.07	±1.0	±0.4	±0.08	±0.03	±0.01	±0.02
Determinantny, 1-node	17.8	8.2	2.22	14.0	5.2	2.71	0.73	0.64	0.13
youngest leaf	±0.8	±0.2	±0.09	±0.2	±0.2	±0.06	±0.03	±0.03	±0.02
Bakara, 6-node	18.2	10.0	1.82	10.3	4.2	2.48	0.51	0.42	0.19
oldest leaf	±0.8	±0.4	±0.02	±1.02	±0.2	±0.01	±0.03	±0.01	±0.03
Bakara, 1-node	16.5	7.5	2.20	9.5	3.5	2.71	0.54	0.47	0.14
youngest leaf	±1.0	±0.0	±0.14	±0.7	±0.0	±0.23	±0.01	±0.00	±0.03

unfolding till the phase of leaf photosynthetic maturity and provoke the corresponding alterations of spectral properties of chloroplasts and leaves.

8.6.4 Development of a Leaf

The process of development of a leaf consists of two main phases – leaf synthesis and leaf degradation [Sestak, 1985]. The first phase is characterized by the ratios $f_m690/f_m735 < 1$ and $f_s690/f_s\,735 < 1$. Values above 1 are a clear sign of leaf degradation (Fig. 8.22). With increasing leaf age and during leaf senescence the *Rfd* values and A_p values decrease (Table 8.11).

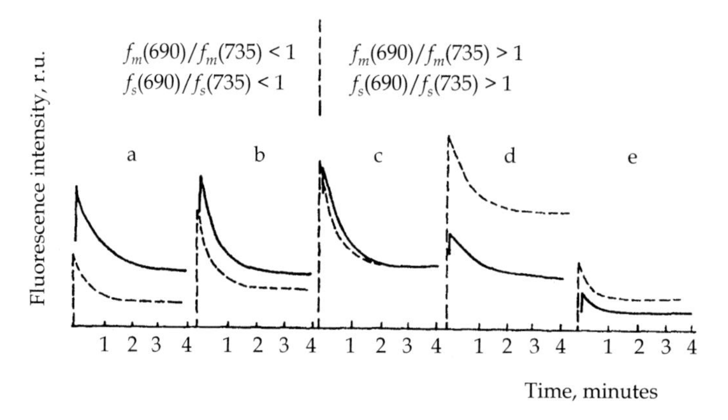

Fig. 8.22 Changes in the chlorophyll fluorescence induction kinetics of dark-adapted leaves of *Pisum sativum* (variety *Determinantny*) during leaf development (*a, b, c*) and degradation (*d, e*). *a* – fully green, *b* – light green, *c* – green-yellowish, *d* – yellow-greenish, *e* – yellow leaves; — 735 nm, - - - 690 nm. Fluorescence excitation: He:Ne-laser, 632.8 nm. The leaves were darkened for 15 minutes prior to illumination [Posudin, 1997, 2002]

The process of leaf development can provoke the changes of chlorophyll fluorescence also. The first stage of leaf development is accompanied by the increase of chlorophyll content. During leaf senescence, the photosynthetic quantum conversion decline, chlorophyll is broken down and the chlorophyll content per leaf area unit decreases. The process of leaf senescence is caused by factors such as progressive mineral eficiency, enhanced water and light stress, air pollution, virus infection and injury by insects [Lichtenthaler and Rinderle, 1988].

Changes in the fluorescence induction kinetics of the leaves under stress conditions (leaf abscission) can be explained by the hydration of the cytoplasma and the chloroplast stroma [Lichtenthaler and Rinderle,

Table 8.11 Effect of senescence of the leaves on the fluorescence parameters of two varieties of *Pisum sativum*. Changes in the chlorophyll fluorescence parameters of detached leaves (duration t of water deficit) in the 690 and 735 nm region.

Variety, Time	*735 nm*			*690 nm*			*Indices*		
	f_d	f_s	*Rfd'*	f_d	f_s	*Rfd"*	$f_m(690)/f_m(735)$	$f_s(690)/f_s(735)$	A_p
Determinantny, $t = 0$	18.30	10.20	1.89	11.00	4.50	2.44	0.54	0.44	0.24
	±0.20	±0.20	±0.03	±0.40	±0.00	±0.09	±0.01	±0.01	±0.20
Determinantny, $t = 4$ hours,	17.50	10.00	1.85	11.00	4.50	2.44	0.54	0.45	0.29
control	±0.20	±0.20	±0.03	±0.40	±0.00	±0.09	±0.01	±0.01	±0.20
Determinantny, $t = 4$ hours	16.50	9.50	1.70	10.50	4.80	2.18	0.58	0.50	0.15
	±0.70	±0.40	±0.06	±0.40	±0.20	±0.13	±0.02	±0.02	±0.02
Determinantny, $t = 24$	18.00	10.00	1.80	10.50	4.50	2.33	0.53	0.45	0.16
hours, control	±0.20	±0.20	±0.03	±0.40	±0.00	±0.09	±0.01	±0.01	±0.20
Determinantny,	3.00	7.80	0.37	1.70	3.00	0.54	0.43	0.31	0.11
t = 24 hours	±0.70	±1.20	±0.07	±0.50	±0.40	±0.11	±0.02	±0.07	±0.04
Bakara, $t = 0$	16.80	9.50	1.77	9.30	4.00	2.33	0.51	0.42	0.17
	±0.60	±0.40	±0.06	±0.90	±0.40	±0.12	±0.03	±0.02	±0.01
Bakara, 4 hours,	17.50	9.50	1.84	10.00	4.00	2.50	0.52	0.42	0.19
control	±0.60	±0.40	±0.06	±0.90	±0.40	±0.12	±0.03	±0.02	±0.01
Bakara, $t = 4$ hours	12.10	7.30	1.66	7.00	3.20	2.22	0.51	0.43	0.17
	±2.00	±1.00	±0.10	±1.40	±0.60	±0.16	±0.06	±0.03	±0.01
Bakara, 24 hours,	17.00	10.00	1.70	10.00	4.50	2.22	0.54	0.45	0.16
control	±0.60	±0.40	±0.06	±0.90	±0.40	±0.12	±0.03	±0.02	±0.01
Bakara, $t = 24$ hours	1.00	5.50	0.18	1.00	2.50	0.40	0.54	0.45	0.16
	±0.60	±1.30	±0.06	±0.50	±0.40	±0.15	±0.06	±0.03	±0.02

1988]. This water deficit provokes the desiccation of a cell and lowers the level of chlorophyll fluorescence emission because of reabsorption of the emitted fluorescence. Besides, the increasing dehydration of leaves causes the changes of the reflection properies of the leaves. Under stress conditions (leaf abscission) the leaf loses the photosynthetic function due to damage of the photosynthetic apparatus. It is shown [Lichtenthaler, 1990] that leaves with lower values of stress-daptation index A_p suffer from stress earlier than those with a higher A_p-value.

8.6.5 Upper and Lower Leaf Sides

The results of chlorophyll fluorescence measurements from upper and lower leaf sides are presented in Table 8.12.

It is shown that the values for the ratios f_m690/f_m735 and f_s690/f_s735, *Rfd* values in 690 nm and 735 nm regions at the lower leaf side exceed the values at the upper leaf side for both species.

The dependence of chlorophyll fluorescence on the leaf side is related to the cross-sectional leaf structure. This structure in a typical dicotyledonous (C_3) plant like a green pea (*Pisum sativum*) consists of the upper and lower epidermis which are separated by the mesophyll; the latter includes the palisade parenchyma cells and the spongy parenchyma cells (spongy mesophyll) [Gausman, 1985]. There is a difference in the structure and cell arrangement of the upper and lower sides of the leaves of C_3 plants – the upper side is characterized by more densely packed palisade parenchyma cells, and higher chlorophyll content and density than the lower one. That is why the re-absorption of emitted chlorophyll fluorescence of the lower side is much lower and fluorescence intensity is higher than in the upper side of the leaf. The fluorescence intensity near 690 nm is higher than in the 735 nm region because of the overlapping absorption and fluorescence spectra at 640–740 nm part of the spectrum.

8.7 VECTOR METHOD OF FLUORESCENCE ANALYSIS OF PLANTS UNDER STRESS CONDITIONS

8.7.1 Introduction

One of the main problems of biomonitoring is related to the search for effective test-objects and test-functions for the determination and estimation of the toxicity of different pollutants in aquatic media. It is possible to use bacteria, algae, protozoa, invertebrates, fish, and plants as test-objects during biomonitoring. Different test-functions such as biomass, growth-rate, photosynthesis, fluorescence, etc., are examined during toxicity testing of plants.

Table 8.12 Effect of leaf side on the fluorescence parameters of two varieties of *Pisum sativum*

Side	735 nm			690 nm			Indices		
	f_d	F_s	*Rfd'*	f_d	f_s	*Rfd"*	$f_m(690)/f_m(735)$	$f_s(690)/f_s(735)$	A_p
Determinantny, upper	17.00 ±1.00	10.70 ±0.20	1.59 ±0.01	10.70 ±0.80	5.00 ±0.40	2.10 ±0.05	0.57 ±0.03	0.47 ±0.04	0.17 ±0.02
Determinantny, lower	14.00 ±0.40	10.00 ±0.40	1.40 ±0.10	11.50 ±0.00	5.70 ±0.20	2.04 ±0.08	0.72 ±0.08	0.57 ±0.01	0.21 ±0.04
Bakara, upper	17.80 ±1.00	9.30 ±0.50	1.91 ±0.14	10.00 ±1.40	3.70 ±0.20	2.72 ±0.35	0.50 ±0.03	0.39 ±0.05	0.22 ±0.02
Bakara, lower	15.70 ±1.00	8.80 ±0.20	1.77 ±0.20	11.10 ±0.80	5.00 ±0.00	2.23 ±0.17	0.66 ±0.01	0.57 ±0.01	0.16 ±0.09

Nondestructive spectroscopic analysis of plants can be realized on the basis of two methodological approaches. The first one provides the examination of fluorescence emission spectra of green leaves which exhibit fluorescence maxima near 440–450 nm (blue region), at 690 nm and at 735 nm (red region). The ratio of the fluorescence intensities *F*690/*F*735, which depends strongly on the chlorophyll content of the leaf, can be used as the fluorescence index that correlates with the rate of photosynthesis and the effect of different stress factors on photosynthetic activity of leaves.

The second approach is based on the detection of fluorescence kinetics of dark-adapted leaves (induction of fluorescence, Kautsky effect). Fluorescence kinetics reflects the sum total of processes which are linked with photosynthesis activity of a plant and also depends on the effect of different stress factors. The measurements of several chlorophyll fluorescence parameters such as vitality index $Rfd = f_d/f_s$, particularly, *Rfd*(735) and *Rfd*(690), and stress adaptation index $A_p = 1 - [Rfd(735)+1]/[Rfd(690)+1]$ provide useful information about healthy status of the leaf under stress conditions.

The vector method of quantitative estimation of stress factors on fluorescence parameters of chlorophyll was proposed for the first time by Posudin (1992). This method is based on excitation of chlorophyll fluorescence, simultaneous control of several fluorescence indices, and estimation of value and direction of the vector P_R, which is constructed in *N*-dimensional system of coordinates (*N* is a number of fluorescence indices which are recorded simultaneously); each axis of this system presents the ratio of fluorescence index under stress conditions and control fluorescence index. It is shown that the value and direction of the vector depend strongly on the status of the plant and the effect of different stress factors on it.

8.7.2 Principle of Vector Method

In the two-dimensional system of coordinates ($N = 2$), the value r and the direction θ are defined as follows:

$$r = \sqrt{(F_1/F_{1c})} + (F_2/F_{2c})^2 \qquad \text{(Eqn. 8.1)}$$

$$\theta = arctan\ [(F_1/F_{1c})/(F_2/F_{2c})] \qquad \text{(Eqn. 8.2)}$$

where F_i = fluorescence index of the sample in stress conditions; F_c = fluorescence index of control sample.

In the three-dimensional system of coordinates ($N = 3$), the value r and the directions (θ_1 and θ_2) are determined as follows:

$$r = \sqrt{(F_1/F_{1c})^2 + (F_2/F_{2c})^2 + (F_3/F_{3c})^2} \quad \text{(Eqn. 8.3)}$$

$$\theta_1 = arccos\ [F_1/F_{1c})/r] \quad \text{(Eqn. 8.4)}$$

$$\theta_2 = arctan\ [(F_3/F_{3c})/(F_1/F_{1c})] \quad \text{(Eqn. 8.5)}$$

The value r and the direction (θ_1, θ_2 and θ_3) of the vector $\vec{R}$ in the four-dimensional system of coordinates ($N = 4$) are determined from the following equations:

$$r = \sqrt{(F_1/F_{1c})^2 + (F_2/F_{2c})^2 + (F_3/F_{3c})^2 + (F_4/F_{4c})^2} \quad \text{(Eqn. 8.6)}$$

$$F_1/F_{1c} = r\cos\theta_1 \quad \text{(Eqn. 8.7)}$$

$$F_2/F_{2c} = r\sin\theta_1\cos\theta_2 \quad \text{(Eqn. 8.8)}$$

$$F_3/F_{3c} = r\sin\theta_1\sin\theta_2\cos\theta_3 \quad \text{(Eqn. 8.9)}$$

$$F_4/F_{4c} = r\sin\theta_1\sin\theta_2\sin\theta_3 \quad \text{(Eqn. 8.10)}$$

The N-dimensional system of coordinates can also be used, but in such situations ($N > 4$) the graphical description of vector $\vec{R}$ is impossible; its value and direction can only be tabulated.

8.7.3 Application of Vector Method

The concrete examples of application of vector method on the basis of fluorescence indices such as *Rfd*(690), *Rfd*(735) and A_p (method of fluorescence kinetics), and *F(690)/F(735)* (method of spectrofluorometry) are detailed below:

1. Registration of two fluorescence indices *Rfd(690)* and *Rfd(735)* of the leaves of green pea *Pisum sativum* of different chlorophyll content:

Value r and direction θ of vector $\vec{R}$ in the two-dimensional system of coordinates that are determined by the formulae (8.1)–(8.2), are given in Table 8.13.

Table 8.13 Relative values of fluorescence indices, value r and direction θ of vector $\vec{R}$ for leaf of green pea *Pisum sativum* of different chlorophyll content

Colour	*Rfd(690)/ Rfd(690)$_c$*	*Rfd(735)/ Rfd(735)$_c$*	*r*	*θ,°*
Dark green (control)	1.00	1.00	1.41	45.00
Green	0.70	0.80	1.06	41.20
Light green	0.54	0.65	0.84	39.70
Yellow green	0.30	0.37	0.48	39.00

The dependence of value r and direction θ of vector $\vec{R}$ in the two-dimensional system of coordinates on the colour of leaves is presented in Fig. 8.23.

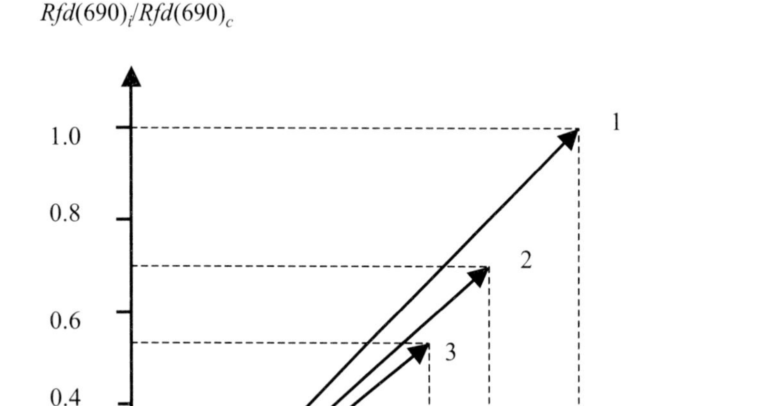

Fig. 8.23 Dependence of value r and direction θ of vector $\vec{R}$ in the two-dimensional system of coordinates on the colour of leaves of green pea *Pisum sativum*: 1 – dark green leaf (control); 2 – green; 3 – light green; 4 – yellow green leaf

2. Registration of three fluorescence indices *Rfd*(690), *Rfd*(735) and A_p for the leaves of green pea *Pisum sativum* of different chlorophyll content:

Value r, direction (θ_1 and θ_2) of vector $\vec{R}$ in the three-dimensional system of coordinates that are determined by the formulae (8.3)–(8.5), are given in Table 8.14.

The dependence of value r and direction (θ_1 and θ_2) of vector in the three-dimensional system of coordinates on the colour of leaves is presented in Fig. 8.24.

3. Registration of two fluorescence indices *Rfd*(690) and *Rfd*(735) of the leaves of green pea *Pisum sativum* under water deficit conditions:

Table 8.14 Relative values of fluorescence indices, value r and direction (θ_1 and θ_2) of vector $\vec{R}$ in the three-dimensional system of coordinates for leaf of green pea *Pisum sativum* of different chlorophyll content

Colour	$Rfd(690)_i/Rfd(690)_c$	$Rfd(735)_i/Rfd(735)_c$	A_{pi}/A_{pc}	r	θ_1, °	θ_2, °
Dark green (control)	1.00	1.00	1.00	1.73	54.68	45.00
Green	0.70	0.80	0.54	1.19	53.96	37.64
Light green	0.54	0.65	0.37	0.92	54.06	34.42
Yellow green	0.30	0.37	0.20	0.52	54.76	33.69

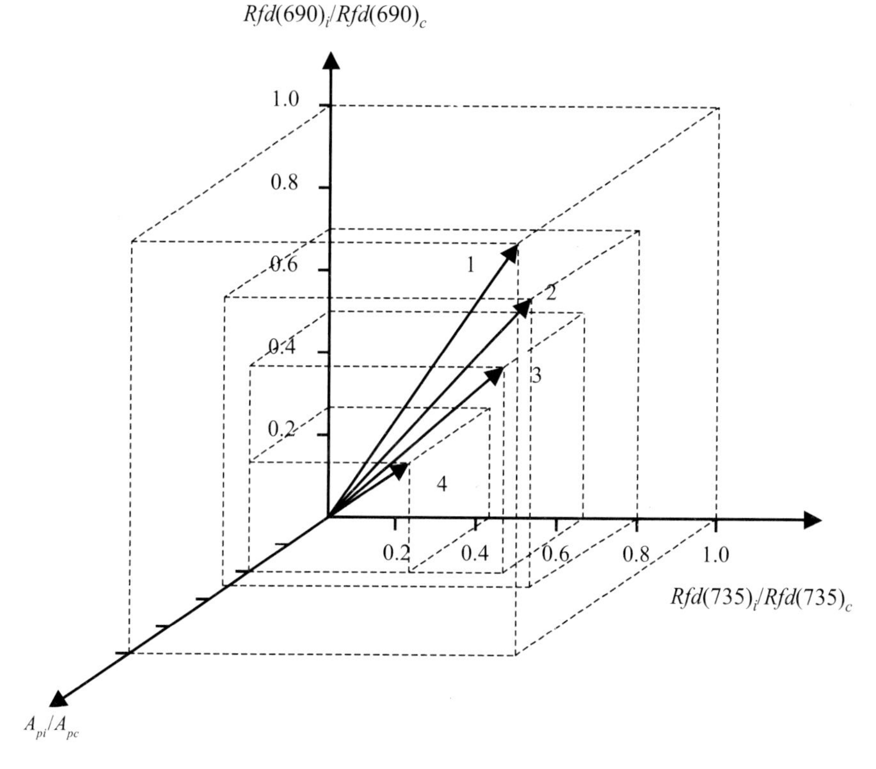

Fig. 8.24 Dependence of value r and direction (θ_1 and θ_2) of vector $\vec{R}$ in the three-dimensional system of coordinates on the colour of leaves of green pea *Pisum sativum*: 1 – dark green leaf (control); 2 – green; 3 – light green; 4 – yellow green leaf

Value r and direction θ of vector $\vec{R}$ in the two-dimensional system of coordinates that are determined by the formulae (8.1)–(8.2), are given in Table 8.15.

Table 8.15 Relative values of fluorescence indices, value r and direction θ of vector $\vec{R}$ for leaf of green pea *Pisum Sativum* under water deficit conditions

Water Deficit, Hours	$Rfd(690)_i/Rfd(690)_c$	$Rfd(735)_i/Rfd(735)_c$	r	$\theta,°$
$t = 0$ (control)	1.00	1.00	1.41	45.0
$t = 1.5$	0.83	0.91	1.23	42.4
$t = 4$	0.57	0.64	0.86	41.7
$t = 24$	0.07	0.09	0.11	37.9

The dependence of value r and direction θ of vector $\vec{R}$ in the two-dimensional system of coordinates on the water deficit of leaves is presented in Fig. 8.25.

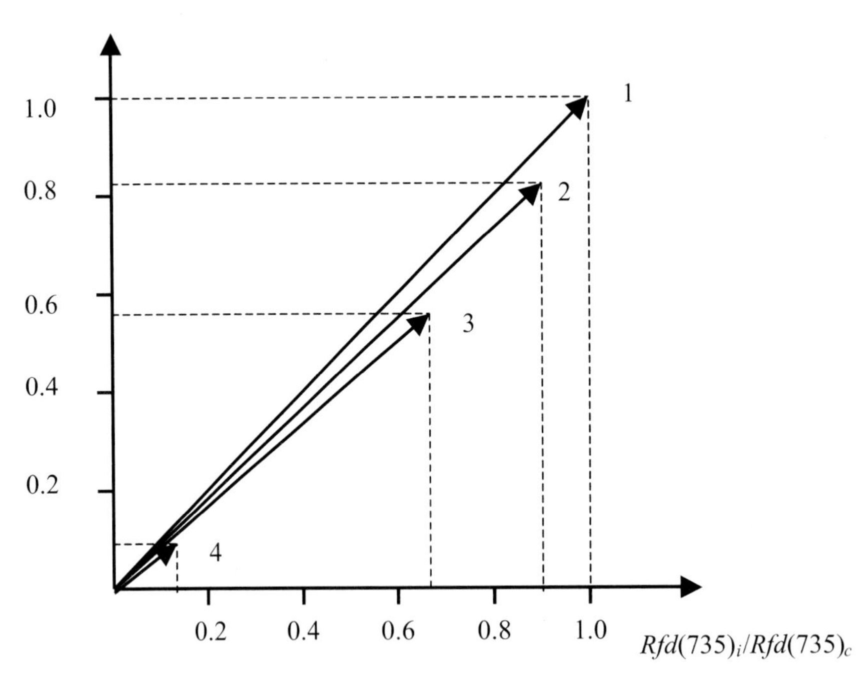

Fig. 8.25 Dependence of value r and direction θ of vector $\vec{R}$ in the two-dimensional system of coordinates on the water deficit of leaves of green pea *Pisum sativum*: 1 – control; 2 – after 1.5 hours of water deficit; 3 – after 4 hours; 4 – after 24 hours

4. Registration of three fluorescence indices $Rfd(690)$, $Rfd(740)$ and A_p for the leaves of green pea *Pisum sativum* under water deficit conditions:

Value *r*, direction (θ_1 and θ_2) of vector $\vec{R}$ in the three-dimensional system of coordinates that are determined by the formulae (8.3)–(8.5), for the leaves of green pea *Pisum sativum* with normal water supply and under water stress are given in Table 8.16.

Table 8.16 Relative values of fluorescence indices, value *r* and direction (θ_1 and θ_2) of vector $\vec{R}$ in the three-dimensional system of coordinates for leaves of green pea *Pisum sativum* with normal water supply and under water stress

Water Deficit, Hours	$Rfd(690)_i/Rfd(690)_c$	$Rfd(740)_i/Rfd(740)_c$	A_{pi}/A_{pc}	r	θ_1,°	θ_2,°
t = 0 (control)	1.00	1.00	1.00	1.73	54.69	45.0
t = 2	0.83	0.79	0.96	1.49	56.15	49.15
t = 4	0.47	0.42	0.92	1.12	65.19	62.94

The dependence of value *r* and direction (θ_1 and θ_2) of vector in the three-dimensional system of coordinates on the water deficit of leaves of green pea *Pisum sativum* is presented in Fig. 8.26.

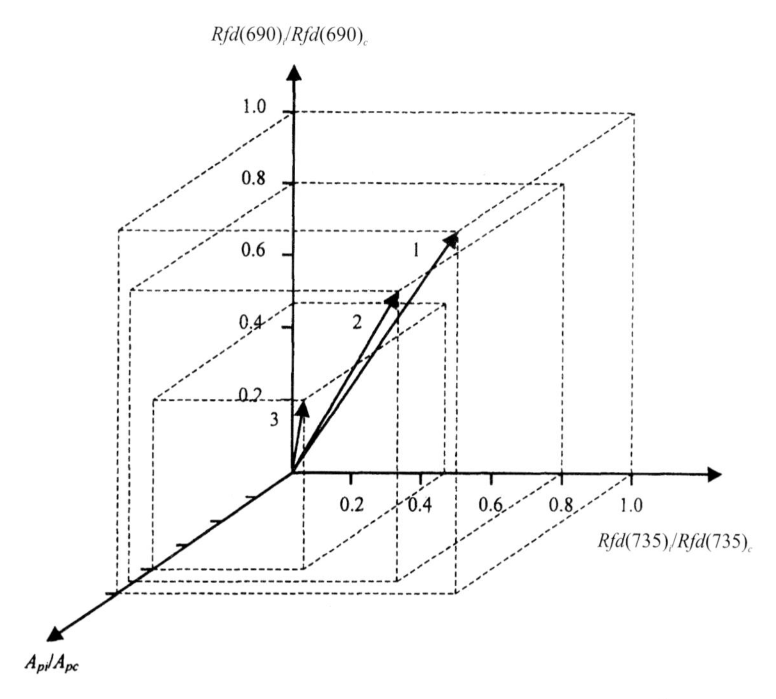

Fig. 8.26 Dependence of value *r* and direction (θ_1 and θ_2) of vector $\vec{R}$ in the three-dimensional system of coordinates on the water deficit of leaves of green pea *Pisum sativum*: 1 – control; 2 – after 2 hours of water deficit; 3 – after 4 hours

5. Registration of four fluorescence indices *Rfd*(690), *Rfd*(740), A_p and *F*690/*F*735 for the leaves of green pea *Pisum sativum* under water deficit conditions:

Value *r*, direction (θ_1, θ_2 and θ_3) of vector in the four-dimensional system of coordinates that are determined by the formulae (8.6)–(8.10), for the leaves of green pea *Pisum sativum* with normal water supply and under water stress are given in Table 8.17.

8.8 SUMMARY

Method of laser spectrofluorometry *in vivo* of the leaves is rather promising for laboratory investigations of the effect of different agrochemical treatments and external physical factors; it is possible to control all the stages of growth and development of the plant, to study the effects of side and segment of the leaf, its nodal position, and the age. In some cases it is possible to achieve the species identification of the plants on the basis of the relevant spectral indices analysis.

The main perspective is the possible application of the results from single leaf investigation to the monitoring of agronomic plants at the canopy level and remote sensing of stressed agronomic fields on the basis of laser-induced fluorescence measurements.

Analysis of fluorescence induction kinetics of chlorophyll provides useful information about the process of the development of a leaf. Formation of internal leaf structure, increase of pigment content and intense fixation of photosynthetic activity accompany the first phase of leaf dvelopment. The next phase of leaf development is characterized by a prevalence of degradative processes such as declination of photosynthetic quantum conversion and decrease of chlorophyll content per leaf area unit. Both these processes provoke the relative changes of spectral properties of the leaf.

It has been found in these experiments that two varieties of the same species of green pea demonstrated certain differences between the fluorescence indices.

The measurements of several chlorophyll fluorescence parameters such as the fluorescence decrease f_d, maximum fluorescence f_m, steady-state fluorescence f_s, ratios f_m690/f_m735, f_s690/f_s735 and *F*690/*F*735, vitality indices *Rfd*′ at 735 nm and *Rfd*″ at 690 nm and stress adaptation index A_p provide useful information about the development of a leaf and its healthy status under stress conditions.

Table 8.17 Relative values of fluorescence indices, value *r* and direction (θ_1, θ_2 and θ_3) of vector $\vec{R}$ in the four-dimensional system of coordinates for leaf of green pea *Pisum sativum* with normal water supply and under water stress condition.

Water Deficit, Hours	$Rfd(690)_i/Rfd(690)_c$	$Rfd(740)_i/Rfd(740)_c$	A_{pi}/A_{pc}	$F(690)/F(740)_i/F(690)/F(740)_c$	r	θ_1,°	θ_2,°	θ_3,°
t = 0 (control)	1.00	1.00	1.00	1.00	2.00	60.00	54.72	45.00
t = 1.5	0.83	0.91	0.75	1.12	1.82	62.87	59.17	32.61
t = 4	0.57	0.64	0.63	1.33	1.70	70.40	69.14	24.90
t = 24	0.07	0.09	0.12	1.37	1.38	87.09	87.09	4.99

It is possible to use some of these fluorescence parameters as taxonomic criteria during investigation of the effect of age and nodal position of the leaves (f_m690/f_m735 and f_s690/f_s735) and of leaf side (f_m690/f_m735, f_s690/f_s735, *Rfd′* and *Rfd″*).

The vector method of quantitative estimation of stress factors on fluorescence parameters of chlorophyll was applied to study the status of the plant and the effect of different stress factors on it. The simultaneous recording of several fluorescence parameters of chlorophyll and application of the vector method of biomonitoring can make it possible to realize the more precise quantitative estimation of these stress factors.

8.9 REFERENCES

Chappelle, E.W., Wood, F.M., McMurtrey, J.E. and Newcomb, W.W. 1984*a*. Laser-induced fluorescence of green plants. 1. A technique for remote detection of plant stress and species differentiation. *Applied Optics* 23: 134–138.

Chappelle, E.W., Wood, F.M., McMurtrey, J.E. and Newcomb, W.W. 1984*b*. Laser-induced fluorescence of green plants. 2. LIF caused by nutrient deficiencies in corn. *Applied Optics* 23: 139–142.

Chappelle, E.W., McMurtey, J.E. III and Kim, M.S. 1991. Identification of the pigment responsible for the blue fluorescence band in the laser-induced fluorescence (LIF) spectra of green vegetation plants, and the potential use of this band in remotely estimating rates of photosynthesis. *Remote Sens. Environ.* 36: 213–218.

Gausman, H.W. 1985. Plant Leaf Optical Properties in Visible and Near-Infrared Light. *Graduate Studies Texas Tech University*, 29: 78.

Gorodny, N.M., Povkhan, M.F., Posudin Yu.I. and Bykin, A.V. 1992. Fluorescence methods of control of the fertilizers efficiency. *Agrochemistry*. 4: 133–138.

Goulas, Y., Moya, I. and Schmuck, G. 1990. Time-resolved spectroscopy of the blue fluorescence of spinach leaves. *Photosynth. Res.* 25: 299–307.

Goulas, Y., Moya I., and Schmuck, G. 1991. Time-resolved spectroscopy of spinach leaves. In: *Proc. 5th Intern. Colloq. Phys. Measurements and Signatures in Remote Sensing*. Courchevel, pp. 715–718.

Hall, D. 1983. *Photosynthesis*. Moscow, Mir. p 132.

Ivanitskaya, S.A., and Posudin, Yu.I. 1989. Laser spectrofluoremetry of cereals. Biol. Sciences, Depon. VINITI, N5350-B89, p 10.

Ivanitskaya, S.A. and Posudin, Yu.I. 1989. Laser spectrofluorometry of vegetables. Biol. Sciences, Depon. VINITI, N5351-B89, p 9.

Kautsky, H. and Franck, U. 1943. Chlorophyll fluorescenz und kohleusäure assimilation. *Biochemische Zeitschrift*. 315: 139–232.

Kautsky, H. and Hirsch, A. 1931. Das Fluoreszenzverhalten gruner Pflanzen. *Biochem. Z.* 274: 422.

Kautsky, H. and Hirsch, A. 1931. Neue Versuche zur Kohlenstoffassimilation. *Naturwissenschaften*. 19: 964.

Kays, S.J. and Paull, R.E. 2004. *Postharvest Biology*. Exon Press, Athens. p 568.

Kochubey, S.M. 1986. *Organization of pigments of photosynthetic membranes as the basis of energy supply of photosynthesis*. Kiev: Naukova Dumka, p 192.

Lang, M., Stober, F. and Lichtenthaler, H.K. 1991. Fluorescence emission spectra of plant leaves and plant constituents. *Padiat. Environ. Biophys.* 30: 333–347.

Lang, M., Siffel, P., Braunova, Z. and Lichtenthaler, H.K. 1992. Investigation of the Blue-green Fluorescence Emission of Plant Leaves. *Bot. Acta.* 105: 435–440.

Lichtenthaler, H.K. 1987. Chlorophylls and carotenoids, the pigments of photosynthetic biomembranes. *Methods Enzymol.* 148: 350–382.

Lichtenthaler, H.K. and Rinderle, U. 1988. The role of chlorophyll fluorescence in the detection of stress conditions in plants. *CRC Critical Reviews in Analytic Chemistry.* 19: 29–85.

Lichtenthaler, H.K. 1988*b*. In *vivo* chlorophyll fluorescence as a tool for stress detection in plants. In: *Application of Chlorophyll Fluorescence in Photosynthesis Research, Stress Physiology, Hydrobiology and Remote Sensing* (Lichtenthaler H.K., ed.), Kluwer Academic Publishers, Dordrecht, pp. 129–142.

Lichtenthaler, H.K., Burkard, G., Kuhn, G. and Prenzel, U. 1981. Light-induced accumulation and stability of chlorophylls and chlorophyll proteins during chloroplast development in radish seedlings. *Z. Naturforsh*, 36: 421–430.

Lichtenthaler, H.K. 1990. Applications of chlorophyll fluorescence in stress physiology and remote sensing. In: *Applications of Remote Sensing in Agriculture* (Steven, M. and Clark, J.A., eds.) Butterworths Scientific Ltd., London, pp. 287–305.

Lichtenthaler, H.K., Lang, M. and Stober, F. 1991. Laser-induced blue and red chlorophyll fluorescence signatures of differently pigmented leaves. In: *Proc. 5th Intern. Colloq. Phys. Measurements and Signatures in Remote Sensing.* Courchevel. pp. 727–730.

Lichtenthaler, H.K., Stober, F. and Lang, M., 1993. Laser-induced fluorescence emission signatures and spectral fluorescence ratios of terrestrial vegetation. In: *Intern. Geoscience and Remote Sensing Symposium, IGARSS' 93*, Tokyo, Japan. V.III, pp. 1317–1320.

Litvin, F.F. and Belyaeva, O.B. Characteristics of certain reactions and general scheme of biosynthesis of native forms of chlorophyll in etiolated plant leaves. *Biochemistry* (Russian) 36: 615–622.

Posudin, Yu.I. Instrument for measurement of fluorescence decay. Invention N 1263039 USSR, 1986.

Posudin Yu. I. 1988. *Lasers in Agriculture.* Science Publishers, Inc: Enfield, NH, USA. p 188.

Posudin Yu.I. 1992. *Optical Methods of Investigation of Photobiological Reactions of Higher and Lower Plants.* Thesis for a Doctor's degree. St. Petersburg: Agrophysical Institute, 1992. p 433.

Posudin, Yu.I. 1996. *Methods of Optical and Laser Spectroscopy in Agriculture.* Seminar Series of visiting Fulbright Scholar. University of Georgia, USA. January-April 1996.

Posudin, Yu.I. 1997. Spectroscopic monitoring of agronomic plants and canopies. *Bull. Nat. Agr. Univ.* 1: 14–21.

Posudin, Yu.I. and Kozhemjako, Y.V. 2001. Method of spectroscopic analysis of plants under stress conditions. *Bull. Uzhgorod Nat. Univ.* 10: 120–123.

Posudin, Yu.I. 2002. Fluorescence analysis of plants during development under stress conditions. *Bull. Nat. Agr. Univ.* 58: 42–48.

Posudin, Yu.I. and Kozhemjako, Y.V. 2002. Fluorometer for analysis of agronomic plants under stress conditions. *Bull. Nat. Agr. Univ.* 58: 39–41.

Posudin, Yu.I. and Kozhemyako, Y.V. 2003. Method of Analysis of Plants under Stress Conditions. Patent of Ukraine, 61148, N2001042240, 2003.

Posudin, Yu.I. 2003. Vectorial method of fluorescence analysis of plants under stress conditions. *Bull. Nat. Agr. Univ.* 63: 45–49.

Posudin, Yu.I., Kucherov, A.P. and Olkhovskaya, G.P. 1991. Decomposition of spectral curves. Biol. Sciences, N254-B91, p 14.

Posudin, Yu.I., Ivonchik, P.N., Ivanitskaya, S.A. and Lilitskaya, G.G. 1992. Physical and biological methods of pesticides control. *Bull. Agr. Science.* 12: 36–38.

Sestak, Z. 1985. *Photosynthesis during leaf development.* Academia, Praha, Dr. W. Junk, Dordrecht-Boston-Lancaste.

Sestak, Z. and Siffel, P. 1988. Changes in chloroplast fluorescence during leaf development. In: *Application of Chlorophyll Fluorescence in Photosynthesis Research, Stress Physiology, Hydrobiology and Remote Sensing* (Lichtenthaler H.K., ed.), Kluwer Academic Publishers, Dordrecht, pp. 85–91.

Stober, F. and Lichtenthaler, H.K. 1992. Changes of the laser-induced blue, green and red fluorescence signatures during greening of etiolated leaves of wheat. *J. Plant. Physiol.* 140: 673–680.

Stober, F. and Lichtenthaler, H.K. 1993. Studies of the localization and spectral characteristics of the fluorescence emission of differently pigmented wheat leaves. *Bot. Acta,* 106: 365–370.

Stober, F., Lang, M. and Lichtenthaler, H.K. 1994. Blue, Green, and Red Fluorescence Emission Signatures of Green, Etiolated, and White Leaves. *Remote Sens. Environ.* 47: 65–71.

Subhash, N. and Mohanan, C.N. 1994. Laser-induced red chlorophyll fluorescence signatures as nutrient stress indicator in rice plants. *Remote Sens. Environ.* 47: 45–50.

Conclusion

Present-day agricultural industry requires the development of new effective methods of quality evaluation of agricultural production. Methods of optical and laser spectroscopy are based on the interaction of optical radiation with the matter.

Such types of interaction can be distinguished as: (1) *absorption* – the process in which incident radiated energy is retained without reflection or transmission on passing through a medium; (2) *reflection* – the phenomenon of a propagating light wave being thrown back from a surface; (3) *luminescence* (*fluorescence*) – which occurs when the energy source transfers an electron of an atom from a lower (ground) energy state into an excited higher energy state, then the electron releases the energy in the form of light when it returns back to a lower energy state; (4) *light scattering* – the change of some characteristics of the light flow (space distribution of light intensity, frequency spectrum, polarization) during its interaction with the matter. All spectroscopic methods offer non-destructive, rapid and precise evaluation of agricultural products.

Method of absorption/transmission spectroscopy of milk in the visible part of the spectrum demonstrates a serious limitation – a milk sample must be adequately diluted to obtain the linear dependence of optical density on the concentration of milk sample, which is to be evaluated.

Method of infrared microspectrophotometry of milk is based on the measurement of the absorption of the main components of milk (fat, protein, lactose) and quantitative evaluation of these components. The spectral position of infrared absorption bands depends on the temperature, level of milk homogenization, and pH of the milk.

A number of milk components (proteins, fat, vitamins) demonstrate fluorescence. However, the fluorescence intensity of milk depends linearly on the concentration of the sample for diluted solutions only; the fluorescence method can be used in laboratory measurements.

Method of near-infrared spectroscopy provides the possibility to analyze milk samples which contain a high proportion of water and demonstrate high level of opacity. The NIR method is rapid, non-destructive and offers a high level of accuracy in comparison with traditional chemical methods.

Method of light scattering can be used for analysis of the size distribution of milk particles and the effect of technological processes which are related to heating (pasteurization, sterilization and homogenization) during this distribution.

Spectroscopic methods can be applied to non-destructive evaluation of poultry-farming products. The intensity of reflected light is different for white and brown eggshells. The spectral position of maximum transmission of eggs depends on the freshness of the eggs and presence of technical spoilage. The intensity of fluorescence maxima depends on the freshness of eggs, level of eggshell pigmentation, and interior defects. Laser radiation with small divergence can be applied for the detection of eggshell cracks.

Laser radiation can interact with a bird's feather which presents a periodical structure and produces a diffraction pattern. The relative intensity and spatial position of diffraction maxima depends on the geometry of the feather (diameter of the ramus and radius) which reflects the healthy status of the bird, diseases, and regime of feeding.

Spectroscopic methods can be explored in honey-breeding. All types of interaction of optical radiation with honey can be used, in principle, for fast and precise diagnostics of this product. The chemical composition, physical properties, geographic origin and age of honey are closely related to its spectral parameters which can be used as taxonomic indices or indicators of honey state and quality.

The near-infrared reflectance spectroscopy gives useful information about the adulteration of honey.

The dependence of transmittance of horse hair on the illuminance was investigated in these experiments. It was shown that the transmittance vs. illuminance curves depend strongly on the colour and part of the animal body; each curve is characterized by a linear part and the region of saturation. The linear dependence of transmittance on the illuminance can be used for quantitative estimation of colour of hair-covering of horses.

The fluorescence intensity of the hair depends on the part and colour of the body (sportive horses, ponies and Przevalsky horses). Method of microspectrofluorometry of horse hair can be applied for quantitative evaluation of the level and character of pigment distribution within hair, studying internal structure of hair, diagnostics of state of the horse under influence of feeding conditions and effects of external factors.

It is very informative to use fluorescence parameters of animal (ponies and Przevalsky horses) hair as genotypic signs of animal diversity for determining the degree of heredity.

Spectroscopic methods are widely adopted in agronomic plant-growing. Chlorophyll is the main plant pigment which can fluoresce. The method of laser spectrofluorometry of plant chlorophyll *in vivo* offers useful information about the growth stage, illumination regime of the plant, effect of soil composition and diseases of the plant. Analysis of temporal dependence of the fluorescence intensity of chlorophyll (so-called induction of fluorescence), particularly the measurement of several chlorophyll fluorescence parameters such as the fluorescence decrease f_d, maximum fluorescence f_m, steady-state fluorescence f_s, ratios f_m690/f_m735, f_s690/f_s735 and $F690/F735$, vitality indices Rfd' at 735 nm and Rfd'' at 690 nm, and stress adaptation index A_p provides useful information about the development of a leaf: the first phase is accompanied by the formation of internal leaf structure, increase of pigment content and intense fixation of photosynthetic activity, while the next phase of leaf development is characterized by a prevalence of degradative processes such as declination of photosynthetic quantum conversion and decrease of chlorophyll content per leaf area unit. Both these processes provoke the relative changes of spectral properties of the leaf. In addition, the method of induction of fluorescence is very sensitive to the effects of external stress conditions on the plants. It is also possible to use some fluorescence parameters as taxonomic criteria, during investigation of the effect of age and nodal position of the leaves (f_m690 / f_m735 and f_s690/f_s735) and of leaf side (f_m690 /f_m735, f_s690/f_s735, Rfd' and Rfd'').

In this way, all the above-mentioned spectroscopic methods offer non-destructive, rapid and precise evaluation of the state and quality of agricultural products.

Index